Ecorestoration of Degraded Hills

Edited By

B.K. Tiwari

and

Surendra Singh

Kaushal Publications

Shillong - 793 014

1995

First Published in India,
by

Kaushal Publications,
Dohsingthiang, Nongshiliang,
Nongthymmai, Shillong-793014.

Edition : 1995

ISBN : 81-900447-1-0

Price : Rs. 300 (US $ 60.00)

Typeset by : Printografix, 8/4 Prem Nagar, Janakpuri, New Delhi-110058.
Printed by : Deepak Offset Printers, Vishwas Nagar, Delhi-110032.

Preface

Interaction of man with the environment is inseparable as man derives all his requirements from the biotic and abiotic components of the environment around him. The civilization and cultural traits are important parts of human environment. The modern culture, which is largely based on the materialism of the West and follows limitless principles of economic development, does not fit under the ambit of ecological jurisprudence which impose the limits to development and requires to follow some natural laws that regulate the relationships of organism with environment. The habitat dictates these rules of the environmental limits and defines the *niche* of an organism including man in the system. Thus, there seems to be a conflict between nature, controlled environmental systems and uncontrolled human activities which disturb the interrelationships among animate and inanimate components of a particular ecosystem. The key component of ecorestoration is to prepare an integrated strategy for rehabilitation of an ecosystem which is dynamic and where the equilibrium between regeneration and utilization of resource is maintained. The systems which are studied under ecorestoration of an ecosystem where man is the key component involve the decision-making processes, its social acceptance and also its short and long term effects on the activities of man, the resources and functioning of the system.

The question arises how to accelerate decision-making processes to restore the degraded environment? Say for example, the questions of resource utilization, environmental degradation, pollution and the qualitative improvement of life and so on deal with ecorestoration and self-sustained growth of the system. The issues and spot-seen-problems of ecodevelopment and ecorestoration of any system are closely related to three major aspects :

(a) to search for major issues related to environmental degradation,

(b) to highlight the problems of local inhabitants related to other abiotic and biotic factors of the environment, and

(c) to provide solution for optimal growth and attainment of balance between use and replenishment of resources.

There is ample need to stress on these issues and problems of ecodevelopment especially with reference to the environmental conditions prevailing in the hill areas of North-East India where human interaction with nature is intensified and is adversely affecting the rich and well-established intricate balance between man and environment. For studying the issues and problems of eco-development, the Meghalaya plateau of the North-Eastern Hill region of the country is chosen as a case study. The plateau is provided with strong geo-processes and harbours rich biological resources required for supporting a sustainable society. Despite of all favorable environmental conditions, the forests, the aquatic bodies, the wildlife, and the landuse system of the region are greatly disturbed. The Meghalaya plateau is a true representative tropical as well as sub-tropical wet forest lands of the world because it receives the highest rainfall of the world with the moderate temperature. The Cherranpunjee area is the most affected part of the plateau which is now called-'the Tropical Wet Desert'. Denudational forces and processes of degradation accelerating in these areas are to be studied carefully for the restoration of ecosystems.

The present volume is the product of the deliberations in two workshops, namely, '**Environmental Problems of Meghalaya**' and '**Technologies for Ecorestoration for Cherrapunjee Plateau**' organised by the Centre for Eco-Development, North-Eastern Hill University, Shillong, in which the real issues and problems of environmental degradation were discussed involving the local voluntary organisations, villagers and scientists from various research organisations. We are hopeful that, through this volume, the ground realities of the eco-development problems of the hill areas of the region have been highlighted and their solutions are provided for the protection of the environment in an integrated manner. The studies included in this volume have wide applications in understanding the environmental problems and regional developmental planning particularly with reference to the hill areas of the tropical and sub-tropical lands of the world.

Place : Shillong — B.K. Tiwari

Date : 25th January 1995. — S. Singh

List of Contributors

1. Dr. S. K. Kar, Geologist, Geological Survey of India, Shillong.

2. Mr. B. S. Corrie, IFS, Dy. Conservator of Forests, Ministry of Environment and Forest, Regional Office, Government of India, Shillong.

3. Dr. D. N. Borthakur, Ex. Vice-Chancellor, Assam Agricultural University, Jorhat.

4. Dr. R. N. Prasad, Director, ICAR Research Complex for North-Eastern Hill Region, Shillong.

5. Dr. Arun Verma, Scientist, Animal Nutrition, ICAR Research Complex for North-Eastern Hill Region, Shillong.

6. Dr. B. K. Tiwari, Head, Centre for Eco-Development, North-Eastern Hill University, Shillong.

7. Dr. C. B. Pandey, Research Associate, Regional Centre for National Aforestation and Eco-Development Board, North-Eastern Hill University, Shillong.

8. Dr. M. L. Khan, Lecturer, Department of Forestry, North-Eastern Regional Institute of Science & Technology, Itanagar.

9. Dr. V. Darlong, Scientist, Ministry of Environment and Forest, Regional Office, Government of India, Shillong.

10. Dr. A. Singh, Scientist, ICAR Research Complex for North-Eastern Hill Region, Shillong.

11. Dr. R. S. Tripathi, Professor, Department of Botany, North-Eastern Hill University, Shillong.

12. Dr. Uma Shankar, Research Associate, Department of Botany, North-Eastern Hill University, Shillong.

13. Dr. H. N. Pandey, Reader, Department of Botany, North-Eastern Hill University, Shillong.

14. Dr. R. K. Rai, Professor, Department of Geography, North-Eastern Hill University, Shillong.

15. Dr. Surendra Singh, Reader, Department of Geography, North-Eastern Hill University, Shillong.

16. Dr. S. K. Shahi, Research Associate, Regional Centre for National Aforestation and Eco-Development Board, North-Eastern Hill University, Shillong

List of Tables

Table No.	Name	Page
2.1	Basic Data Input Chart	11
3.1	Forest Department Setup	18
4.1	Physical and Chemical Characteristics of Soils	42
4.2	Important Species of Sacred Groves and Reserve Forests at Cherranpunjee	43
4.3	Floristic Composition of Five Grassland Communities at Cherrapunjee	45
4.4	Nutrient Concentration in Shoots of Dominant Grasses and Dietary Requirements of Mineral Elements	47
4.5	Annual Uptake and Release and Recycling Coefficents of Six Mineral Elements	48
4:6	Changes in the Chemical Properties of Surface Soils after Fire Effects	49
4.7	Annual Input of Nutrients through Rainfall and their Losses through Run-off and Percolation in three Grassland Communities	50
5.1	Watershed Development in one Hectare	56

5.2 Cost that Vary of one Annual of Easch Enterprise ... 58

5.3 Farm-Benefit from Products and By Products of Each Animal ... 58

5.4 Marginal Rate of Return from Livestock Unit ... 58

5.5 Net Gain for Farmers Every Year ... 59

6.1 Landuse Patterns in Meghalaya ... 64

6.2 No. of Households Involved in Jhum Cultivation ... 65

6.3 Classification of Jhum Land in 1981 ... 66

6.4 Population Involved in Jhum Cultivation ... 67

6.5 Households Invovled in Jhum Cultivation ... 68

6.6 Average Quantity of Crops Produced under Jhum and Permanent Cultivation ... 70

6.7 Households Not Using Improved Seeds and .... Fertilizers ... 72

6.8 Percentage Share of Households Using Improved Seeds, Green Manure, Cowdung and Chemical fertilizers ... 73

6.9 Households Types Practising Jhum, Permanent and Both types of Cultivation ... 74

8.1 Monthly Mean Temperature and Normal Rainfall in Shillong ... 98

8.2 Changing Patterns of Landuse in Meghalaya ... 102

8.3 Area under Forests According to their Legal Status (1984-85) 103

8.4 Changes in Cropping Patterns in Meghalaya 105

8.5 Area Under Various Operational Sizes of Landholdings (1970-71 and 1980-81) 107

8.6 Tenancy Status of Landholdings in Meghalaya (1970-71) 107

8.7 Terms on which Holdings are Wholly taken on Rent 109

10.1 Soil Erosion from Traditional and Forest Landuse Systems 121

10.2 Extent of Awareness about Changes in Environment 122

10.3 Soil Erosion and Rainfall Retention in Scientifically Managed Landuse/Farming Systems 124

11.1 Landuse Changes in the Barapani Area of Meghalaya (1910-11 to 1966-67) 131

13.1 Forests and Sanctuaries under the Control of Government in Meghalaya 150

13.2 Common Mammals and Birds of Meghalaya with their Vernacular Names 151

13.3 New Animal Species/Sub-Species Described form Meghalaya 152

14.1 Density of Survival of Tree Seedlings in Mildy and Highly Distributed Forest in Upper Shillong 164

14.2 Mean Density and Sprouting percentages of Stumps of Four Tree Species 165

14.3 Mean Sprout per Stump in the First Growing Season of Four Tree Species 166

14.4 Survival Rate (in per cent) of the Sprout Emerged from the Stumps in Upper Shillong 166

14.5 Number of Sprouts per Stump in the First Growing Season of Four Tree Species 167

14.6 Survival Rate (in per cent) of the Sprouts Emerged from the Stumps 167

List of Figures

Figure No.	Name	Page
4.1	Variability and Monthly Rainfall and Mean Monthly Maximum and Minimum Temperature	26
4.2	Average Dry Matter Standing State in Above Ground Components in Three Soil Layers at Net Primary Productivity Levels	30
4.3 A	Soil Profile (Foreground)	38
4.3 B	Vast Extension of Greenland and Scattered Trees Growing Alongwith the Streams	38
8.1	Relief Features of Mehghalaya Plateau	99
8.2	Changing Landuse Patterns in Meghalaya	102
8.3	Size of Operational Landholdings in Meghalaya	108

Contents

Preface iii

List of Contributors vii

List of Tables ix

List of Figures xiii

Contents xv

Section-A : Ecorestoration : Concepts and Approaches.

1. Ecorestoration of Cherrapunjee Plateau 3
2. Ecorestoration of Cherrapunjee - A Policy 9
3. Ecorestoration Technologies - A Basic Approach 13
4. Present Status & Strategies for Ecorestoration of Degraded Eco-systems at Cherrapunjee 23

Section-B : Problems of Hill Farming.

5. Livestock Farming in the Areas of Shifting Cultivation 53
6. Jhum Cultivation in Meghalaya 61
7. Land Ownership Pattern in Meghalaya 79
8. Landuse Patterns in Meghalaya 95
9. Experiences of ICAR for Suggesting Improvements to Shifting Cultivation 111

Section-C: Geoenvironmental Degradation.

10. Environmental Degradation and its Control 119
11. Geomorphic Processes and Environmental Degradation 127
12. Management of Hill Resources 135

Section-D: Forest and Wildlife.

13. Wildlife Preservation and Community Action 147
14. Regeneration Potentials of Tree Species in Meghalaya Forests. 161

Subject Index 169

[illegible] Introduction: Concepts and Applications
[illegible] Chirality [illegible]
[illegible] of Chirality [illegible]
[illegible] Technology [illegible]
[illegible]
[illegible] of Chirality [illegible]
[illegible]
[illegible]
[illegible]
[illegible]
[illegible]
[illegible]
[illegible]
[illegible]
[illegible]
[illegible]
[illegible]
Subject Index ....

Section-A

Eco-Restoration: Concepts and Approaches

1

Eco-Restoration of Cherrapunjee Plateau

D. N. Borthakur
Ex. Vice-Chancellor,
Assam Agricultural University, Jorhat

Physiography

The physiography of the North-Eastern Region has been grouped under three divisions, namely, the Meghalaya Plateau, the North-Eastern Hills and Basin and the Brahmputra Valley. The Meghalaya Plateau occupies 13 per cent of the area of the region. According to the geologists, the Meghalaya Plateau is an Eastward extension of the massive block of Peninsular India. The ancient land, submerged partially in the early ages, was subsequently filled up with alluvium deposited jointly by the Ganga and the Brahmputra. The organic movement was so slow and free bulking that the sedimentary beds retained their horizontal character and gave rise to structural platforms, well developed in the Cherrapunjee area. Cherrapunjee thus stands on a vast structural platform which is built on gently dipping sandstones of cretaceous age. Limestone hills are also found scattered over the plateau. From Cherrapunjee, the plateau slopes vary gently Southwards for about 6 kms and then falls rapidly to the plains.

Present Status

Wastelands may be created due to many reasons. Even culturable land may lie waste due to causes such as growth of deep rooted shrubs/weeds, waterlogging due to lack of drainage, low fertility of soil etc. However, the wasteland around Cherrapunjee is due to land degradation or land going out of cultivation due to soil erosion.

It is of interest to mention here that a report on the conditions of land prevailing in Cherrapunjee in the year 1837 is also available from the report of Griffith (1847). William Griffith visited Cherrapunjee in September 1837. His report on Cherrapunjee is reproduced below :

> Cherra is situated in a plain surrounded in every direction by low rounded hills, except to the East and Southeast, on which side there is a deep ravine, the whole plateau rising considerably towards the North, in the direction of Cherra itself. Ravines exist here and there; it is along these, and the water-courses, that the only woody vegetation is to be found. The rest of the surface is clothed with grasses, of which a number of species exist, they are chiefly Anthropogoneae. Two or three Osbeckias exist; a Tradescantia (T. septem clavata) covers certain patches with its bright blue flowers. Three species of Impatiens, two with bright pink flowers are common. Spathoglottis and Anthogonum occur in the flat rocks, which frequently prevail; Arundinaria is seen every where as well as a Smithia with lotus-like blossoms.... The woody vegetation consists of Berberis, Viburnum, Bucklandia, Cleyera Floribus Fragrantis, Petalis Sepalis Oppositis, Myrsine and many others, too numerous indeed to mention.
>
> The woods, towards Cherra, assume that rounded and very determinate form, which is seen so commonly in some parts of England, Bucks for instance. None of the trees arrive to any great size. The generality are low, rounded and stunted. It is in these, that Quercus, Viburnum, and Pandanus may be seen growing side by side'.

It would thus appear that degradation of the area was recorded even at that time.

Practically, there has been no serious attempt at eco-restoration at Cherrapunjee all these years. In the meantime, the heavy rainfall of the area, rapid waterfalls and swift flowing rivers have done enormous damage to the already deforested area, through erosion, leading to complete degradation of the land. It has been estimated that upto about 40 tonnes of top soil per hectare per year may be lost in absence of soil conservation measures in the Meghalaya. Similarly, as per report of the Planning Commission (1963), alone 3.5 million acres of land had been completely ruined by ravines and that another 5 million acres of land were in the grip of acute erosion in Uttar Pradesh. Shifting cultivation, practised in a large scale in the entire North-Eastern Region, has been responsible for land degradation to a great extent. The Khasi Hills, however, has been relatively free from this harmful practice. Other forms of unscientific landuse, such as 'bun' method of cultivation of course are prevalent. Thus, major causes of land degradation around Cherrapunjee are deforestation, heavy rainfall, ravines, and unscientific landuse and above all lack of any scientific/co-ordinated effort or neglect of the area for a long time. Lack of awareness among the people also contributes to improper attention.

Studies and Surveys

The Government of India constituted, as far back as 1959, a committee to make a survey of land classified as `Other uncultivated land excluding fallow lands' and `fallow lands other than current fallows'. The committee visited many states and classified the wasteland into certain broad categories, such as, waterlogged lands, saline and alkali lands, land infested with thick growth of bushes, char lands and eroded lands. The committee also made certain recommendations for reclamation of such areas, although no specific recommendations were made for such situations as prevailing in this region. As a matter of fact, acidic wastelands formed due to erosion have not been studied at all except a few recent sporadic studies made by scientists of North-Eastern Hill University, and some attempts by the Forest Department of Meghalaya on afforestation.

Yet it may be worth while to mention a few of the recommendations made by the committee constituted by the Government of India, since these may through some light on the directions of approaches to be made in this region. A number of recommendations have been made and various methods have been tried out for reclamation of saline and alkali soils. Out of these, only two recommendations may be considered for the degraded land in this region. These are pasture development and afforestation. Another approach of adding organic material such as farmyard manure, molasses, crop residues, green manuring crops etc. is also worth consideration.

The approach to such problems as followed for Uttar Pradesh is also worth mentioning. A systematic survey of the wasteland was followed by demarcation of the areas into 3 broad classes, namely,

- land suitable for cultivation without any specific treatment,
- land which may be made fit for cultivation after it is properly treated, and
- land which is totally unfit for cultivation.

Specific recommendations were then followed after dividing the land into convenient blocks. Construction of low bunds to hold/retain water was also carried out in specific areas. Similar approach would be necessary for the Cherrapunjee Plateau also.

Realising the problem of soil erosion, due to unscientific land use, over grazing and indiscriminate felling of trees in the hilly areas, the Planning Commission provided the funds for pasture development and afforestation during the Second 5-year plan and onwards.

The experience of the Forest Department of Meghalaya at afforestation of Cherrapunjee is not a very happy one, although some important trends could be observed. It has been observed that some undergrowth in the form of grasses are maintained as long as some trees have been growing. Once deforested it is difficult even to maintain grasses in the area. Plantation and establishment of trees in the area is very difficult. However, it has been possible to establish some trees when wide and deep pits are dug and the hole is filled with soil brought from outside before planting. Such trees do not have normal growth and remain stunted. Even such trees may help in ecorestoration to some extent, provided the fallen leaves etc. can be prevented from being washed away by the rains.

Whatever has been discussed so far is, primarily, history only. It has, however, to be appreciated that we must learn from history and history should make us wiser. That is why the need for such a workshop/training has been felt essential. Cherrapunjee has almost gone out of our hands, but there are many such areas that will also go completely out of our hands, if not properly attended immediately. Pasture development is the most logical and scientific approach in areas where soil depth has been completely reduced due to erosion. Covering of the pits around, channelising drainage, water harvesting, enrollment of humus and gradual development of trees, even by digging pit, as mentioned earlier, will have to be attempted to reverse the process of degradation. Special technology for all these will have to be developed suited to the local situation. Even the varieties of crops will have to be selected carefully. For example, spreading yet bushy grasses may help in retaining the humus, soil and moisture, rather than thin, solitary grasses. It will also be essential to mark out areas around and toprevent any further felling of trees in such areas, through both laws and creation of awareness.

It would be necessary to constitute interdisciplinary teams with experts in the field of soil sciences, forestry, soil and water conservation, agronomy or forage science, ecology and sociology etc. to not only survey but also to study the various aspects of the problems and suggest viable technology as well as draw up feasible schemes. Data on soils, landuse, rainfall, run-off under various situations and effective measures including soil conservation measures will have to be properly studied. Besides execution of programmes for ecorestoration and measures to prevent further degradation in the vulnerable areas, development of awareness among all sections of the people including village heads, public leaders and administrators is essential. The ecology and environment of such areas, denuded of trees and even soil, are bound to be adversely affected that should have its serious consequences on all aspects of survival of living beings including man.

2

Eco-Restoration of Cherrapunjee : A Policy

S. K. Kar
Geologist, Geological Survey of India, Shillong

The ecological imbalances or the environmental degradation around Cherrapunjee whether due to its natural variables like physiographic setting, climatic factors or due to the anthropogenic activities or interaction of the both require analysis. For restoration of the system, an indepth analysis of these parameters is a pre-requisite. The need of the hour is thus a concerted effort of the entire community. The desired goals can be achieved through an integrated approach involving different agencies.

The first step in this direction should be environmental education in order to create awareness among people about their environment.

Environmental Education

The need for environmental education is to create an awareness among the local inhabitants regarding the various hazards that they are facing everyday. Through educational programmes on 'Environment' or 'Ecological Balance', people can be educated regarding each individual's role to save Cherrapunjee from a catastrophy - it is facing or heading for, if not checked or restored now. The immediate concern of the people is towards the air we breathe, the water we use daily and the land which provides food and shelter. The air and water pollution is due to rapid industrialization, mining, emanations from automobiles and unplanned growth of human habitation. These unchecked activities exerts pressure on carrying capacity of land and water resources. The ultimate outcome is reflected in continuous shrinking of forest cover and attendant soil erosion.

It is to this context that mass movement and people's participation are utmost necessary. To motivate people to come forward, the basic education on environmental programmes should be considered the 'key thrust' activity.

Role of Government

A constructive interaction between the state Government and the central Government is necessary for speedy implementation of the suggested programmes. A necessary legislation for 'Eco-restoration' may be enforced. Appropriate steps may be taken by Government of Meghalaya and Government of India in this regard. The Government of Meghalaya may arrange publicity through Block Level Offices with the help of audio-visual aids and print media.

An immediate ban on all degenerating processes operating in the area can be brought into force by Government of Meghalaya - as a primary step towards ecorestoration.

Basic Approach

To generate basic data on various scientific/social/ resource generations and aspects for formulation of ecorestoration programmes, the scientific organisations is and the scientific community as a whole should be involved. The basic data input through various organisations is listed in Table-2.1.

The enlisted activities and data source are the basic information required for formulating the ecorestoration programme in its integrated manner. It is a time to act before we reach the point of no return. The entire activities call for people participation wherein the various social and scientific organisations be the 'players of a football team' and the inhabitants of Cherrapunjee be its 'active supporters'. The game can be played under the motto 'Operation Cherrapunjee - The Tropical Wet Desert'.

Table - 2.1 : Basic Data Input Chart.

Agency	Work Components
(1)	(2)
Geological Survey of India	Geomorphological, Geological and Environmental Impact Assessment Maps.
Meteorological Department	Hydro Meteorological Maps.
Central Ground Water Board: Central Water Commission	Geohydrological Map: Data on Hydrological Cycle i.e., the interrelationship between Evapotranspiration, Precipitation with Surface Run-off Infiltration etc.
Indian Council of Agricultural Research, State Department of Agriculture and Soil Conservation	Pedological Maps showing Distribution and Classification of soil fertility Mapping on Agricultural Landuse, Cropping patterns, Crop Intensity and other related parameters.

Contd. ...

(1)	(2)
Deparment of Forest, Botanical Survey of India	Vegetation Cover Maps and related informations.
Zoological Survey of India	Information on Faunal Assemblages.
Anthropological Survey of India, Archaeological Survey of India	History of Cultural Heritage and its preservation.
State Directorate of Mineral Resources, Town & Country Planning, Pollution Control Board, Departmnet of Industries and Related Govt. Organisations and Private Bodies	Mining Activities, Urbanisation, Industrial Set-up, Pollution Control Measures etc.
Meghalaya Science Society, NGOs and Other Social Bodies	Dissemination and Publicity of the Information involving Common People/Students through Popular Talks, Scientific Exhibition meetings/group discussions etc.
North Eastern Council	Financial Resource Generation.

3

Eco-Restoration Technologies : A Basic Approach

B. S. Corrie
Dy. Conservator of Forests
Ministry of Environment and Forests,
Regional Office ,Government of India, Shillong.

Plants are important to man. Without plants, life would come to a standstill. Plants convert the carbon dioxide into oxygen and also provide us food - these two basic functions of plants are responsible for life on this planet. The importance of plant in protecting and preserving the environment needs no further emphasis. In case of Cherrapunjee, we are more concerned about this latter aspect when we are to consider ecorestoration.

Since plants remain 'fixed', i.e., they do not move-they are to draw their nourishment from the soil on which they grow. Thus, plants are affected by the soil and the climate. To consider ecorestoration of Cherrapunjee, we would first have to take into account certain background informations and geographical facts of the place.

(a) Cherrapunjee is said to be so called from the number of streams in the neighborhood and pungee 'a village' (Khasi).

(b) Cherrapunjee is situated on a flat table land about

5.25 kms long and 3.50 kms broad, dipping abruptly on the front and on both sides and rising behind towards the main range, of which it is spur. The surface is intersected by shallow rocky water courses which are natural drains for the deluge that annually visits it. The Western part is undulated and hilly, the Southern tracts are rocky ridges of limestone and coal, and the Eastern part is flat and stony, broken only by low isolated conical mounds.

(c) Cherrapunjee, though having the heaviest rainfall in the world has vast barren or degraded areas. William Griffith who visited Cherrapunjee as far back as 1835 writes

> 'The country immediately adjoining the cantonment is flat, and there are rounded hillocks detitute of any covering but grasses and a few, half shrubby plants. To the Eastward there is a very deep and beautiful valley, but this does not contain any large trees'.

Even in those days trees were absent or were of small size. Vegetation was chiefly composed of grasses, ferns and compositae. Tree ferns were common. Vegetation in the valley was rich implying that soil was washed down from the hills.

(d) Later Joseph Dalton Hooker, the great environmentalist, who visited Cherrapunjee in June 1950 writes

> ...and there is not a tree and scarcely a shrub to be seen, except occasional clumps of Pandanus. Westward the plateau becomes very hilly, bare and grassy ....'
>
> 'The direct effect of this deluge is to raise the little streams of Cherra about fourteen feet in as many hours, and to inundate the whole flat; from which, however, the natural drainage is so complete, so as to render a tract, which in such a climate and latitude should be clothed with exuberant forest, so sterile, that no tree finds support, and there is no soil for cultivation of any kind whatsoever not even of rice.'

Even a century and half ago, the situation was as bad as it is now - a barren area of grassy growth with an occasional tree of small girth here and there. In general areas of grassy growth can be linked to the periodic occurrence of fire. This is especially true in the North-Eastern Region of the country where the practice of shifting cultivation is significantly marked. Further, the local people rearing cattle would definitely have a selfish interest to obtain the fresh flush of edible growth which is highly palatable for cattle. This can be said to be true in case of Cherrapunjee where there is occurrence of annual fires. From the above, the following points emerge.

(a) The soil depth is poor and in several cases absent.

(b) The severe rains wash away the little soil that is available.

(c) Since plants are 'Fixed' and not able to move their growth and in fact very survival has been affected leading to the process of desertification.

(d) The biotic interference would have to be regulated.

Plan of Action

In view of the above, the following plan of action is suggested.

(a) Arrest the processes of soil erosion by suitable soil conservation strategy.

(b) Afforest the area on scientific lines and ensure steady inflow of resources. These provisions are to be made for maintenance over a longer period; this is imperative.

(c) Protect the area from biotic interference, control of fires and grazing etc. It should be borne in mind that any attempt to ecorestore successfully the area should be approached on scientific lines and with abundant care; the cost of which would be prohibitive. Only if we can reconcile to these two points, it would be meaningful to take up any ecorestoration programme. Further, an integrated holistic approach is called for.

Arresting Soil Erosion

The soil conservation measures should include a mix of (a) natural methods such as turfing etc. and (b) mechanical methods such as contour trenches, contour bunds, terracing, gully plugging etc.

The main problem in this area is the high and concentrated rainfall; hence a downpour can result in serious loss of precious soil (it has been estimated that it takes hundreds of years to build up an inch of soil). This can be done with the help of soil conservation measures such as contour trenches, contour bunding and so on.

In places where soil depth is less such soil may be stabilised by turfing; grass seeds may be broadcasted in such areas. Other lower strata plants available locally may also be tried. The soil conservation measures that may be adopted are : Gully Plugging, Contour Bunds, Contour Terracing, Half-moon Pits for Planting Trees, Contour Wattling, Silt Retention Dams, Grassing Waterways, Turfing (stabilization of loose soil with grasses), Weeding and Placing of Weeds along Contours, and Bamboo Matt Barricades.

Afforestation

Trees are the major primary producers of forest ecosystem. Trees convert the sun's energy to a form of energy that is of use to the various life forms. Some of the major influences of forests on the environment are listed below :

(1) Moderation of local climate,

(2) Reduction of wind velocity,

(3) Interception of precipitation,

(4) Reduction of the impact of raindrops,

(5) Increases percolation of water,

(6) Moderates flow of water,

(7) Protects soil from erosion, and

(8) Improves quality and quantity of water.

Some Interesting Facts

1. World Research Institute (WRI) President James Speth warned

> An elephant is slaughtered every 10 minutes and an acre of tropical forest disappeared every second. If these trends continue, one-fourth of all species on the planet earth will be gone in the lifetime of todays young people'.

Scientists estimate 30 million species inhabitate the earth. Kenters Miller, director of the WRI's Bio-diversity Programme noted that preserving the earth species is vital because perhaps a cure for Cancer or AIDS is among the 12000 plant species disappearing each year.

> 'Our supplies of food, medicine, industrial chemicals, even our air and water depend upon natural substances and natural processes'.

2. *Leuceana Leucocephala* (Subabool) - It coppices and makes an excellent fuelwood -Caloric value 42000 to 46000 average annual increments 30-40 cu.m.

3. More than one-third of the annual consumption of wood is estimated to be in the building construction industry.

The East Khasi Hills

The East Khasi Hills District covers an area of 519600 hectare comprise a number of hills, valleys and rivers. The district is richly endowed with natural resources. Forests in the district cover about 159142 hectares or about 30 per cent of the geographical area. Five hydel projects exist in the district. Interestingly, out of the total forest area of 159142 hectares only 13026 hectare is declared as reserve forest under control of the Government. It is related to Forest Department setup which is given below for the state of Meghalaya (Table -3.1).

Table - 3.1: Forest Department Set-up.

Level			
State		CCF	
Circle	CF(HQ)	CF(WL)	CF(SF)
Division	DFO	DFO	DFO
Range	RO	RO	RO
Section	Forester	Forester	Forester
Beat	Guard	Guard	Guard

The Problems/Suggestions

1. Forests are being depleted fast especially in East Khasi Hills district. The middle-man is making huge profits at the cost of the environment.

2. The forestry sector should be given added importance and drastic policy changes are called for.

3. New techniques and technologies should be implemented aiming at conservation.

4. Legislation such as the Forest Conservation Act be implemented.

5. Since most of the forest area falls under private ownership, cooperation of the people is necessary.

6. Strengthening of the protection mechanism is desirable.

7. Population control is a must.

What You Can Do

1. Reduce consumption and wastage of forest products.

2. Ensure that want on felling does not take place in no case allow clear felling of areas.

3. Help the department implement various legislations -more especially the Forest Conservation Act, 1980.

4. Help prevent forest damage by grazing and fire.

5. Give wide publicity to forest influences so as to create awareness about forests and the need to protect the forests.

Strategies

There are three aspects that should be kept in view while preparing the strategies for afforestation as given below :

I. Species Selection

- Indigenous local species to be preferred over exotics,
- Fast growing species to be preferred over slow growing ones,
- Nitrogen fixing species to be preferred over non-nitrogen fixing species,
- Broad leaved species over coniferous species,
- In large open areas light demander to be planted,
- Fire hardy species to be preferred over fire susceptible species, and
- Prime importance to be given to the species preferred by the locals. The list of suitable tree species is given in Table-3.2.

Table - 3.2 : Suitable Tree Species and their Uses.

Tree Species	Timber	Fuelwood	Fruits	Fodder	Ornamental
Pinus Resiya	+	+	-	-	+
Prunus Cerasoides	-	+	-	+	-
Exbucklandia sps	-	+	-	+	+
Erythryna Suberosa	+	-	-	-	+
Prunus Nepalensis	+	+	+	-	-
Alnus Nepalensis	-	+	-	-	-
Cryptomeria Japonica	+	+	-	-	+
Jacoranda Ovulifolia	-	+	-	-	+
Cupressus sps	+	+	-	-	+
Grevelia Robusta	+	+	-	-	+

II. Raising of Plantation

- Seeds from known genetically superior source only to be used,
- Seeds to be planted timely for raising of seedlings,
- High planting to be done with basketted seedlings having in no case less than 30 cms height,
- Planting to be done timely and under no circumstances late planting to be done,
- Mixed plantations will be preferred over monocultures. Species with different rotation ages to be preferred.
- The number of weedings to be done according to necessity so that precious nutrients are not utilized by weeds

more frequent weedlings to be done by cutting weeds than by up-rooting,

- While weeding the cut weeds may be kept along contours to prevent erosion and to recycle nutrients.
- Manuring to be done wherever essential,
- Organic manures preferred over the fertilizers,
- Wherever required pits to be made and filled with transported soils, and
- Planting to be done in conjunction with soil conservation measures.

III. Protection

- Fire protection be afforded to all plantations raised by providing internal and external fire lines,
- Planting of fire hard species along the periphery and within the plantation so as to avoid continuous stretches of fire prone species,
- Fire watchers to be recruited from among the locals residing in the vicinity,
- A system of rewards to the local people to be developed and in the event of no fire during a season cash rewards to the local community is recommended,
- Fire plans to be prepared for all plantations raised,
- Environmental awareness campaigns to be conducted; this will help prevent fires, and
- Some subsidy system to be developed.

4

Present Status and Strategies for Ecorestoration of Degraded Ecosystem at Cherrapunjee

R. S. Tripathi, Uma Shankar and H. N. Pandey
Department of Botany,
North -Eastern Hill University , Shillong

Introduction

The existence of extremely degraded grassland ecosystems in the highest rainfall area of Meghalaya is an unresolved ecological problem. Some ecologists believe that the grassy landscape at Cherrapunjee has emerged as a consequence of deforestation due to age-old practice of slash-and-burn agriculture locally called as *'Jhum'*. They believe that a floristically rich broad leaved evergreen forest existed in the past over gentle slopping hills which today support vast stretches of grasslands on highly degraded and nutrient depleted soil substratum. Some of the earlier records of rainfall, vegetation and soil, however, suggest that the graminacious vegetation at Cherrapunjee has not developed during recent past, rather climate, topography and geology of the area are responsible for the development of this type of vegetation. Dr. T. Oldham (c.f. Allen 1905) has described the effects of these factors on the vegetation in the following words:

> 'To the denuding force of these heavy and sudden falls of rain is also due the almost total absence of any soil in the flats of the hills near the Southern escarpment. All has, in fact, been washed away, and a thin crop of coarse grass alone finds sustenance on the rocky surface. The marked absence of trees, the growth of which is cut off as with the axe along the edges of every flat, is largely due to the same causes but also largely to the blighting storms which sweep over these hills. On the sloping sides of these valleys trees grow abundantly, and at many places luxuriantly : there they are sheltered, and abundantly supplied with moisture'.

The present day landscape of Cherrapunjee does not differ markedly from the above description and is chracterised by vast tracts of grasslands on highly drained flats and slopes of hills with small isolated patches of crooked forests occurring frequently in lowlying areas such as along the streams and rivulets, and gullies and dykes of hillocks. Because sufficient water and nutrients are available throughout the year and the effect of winter storm is less, tree growth is possible on these places. Besides these natural environmental factors, cutting of trees and frequent burning of grasslands by the local people during late winters, before the onset of monsoon, have also contributed substantially to the degradation of land and vegetation at Cherrapunjee. On account of these activities and hilly terrain major part of rain water is channelised through over land flow causing heavy sediment and nutrient losses. Coal and limestone mining and change in landuse pattern due to unprecedented rise in population have also accelerated the pace of degradation of natural resources of the entire Southern parts of Shillong plateau including Cherrapunjee.

In this article, we should examine the present status of vegetation and soil of Cherrapunjee and discuss the strategies for ecorestoration of the area.

Geographical Location and Climate

Cherrapunjee (25^0 15' N latitudes and 91^0 45'E longitudes with 1313 m altitudes from mean sea level) is located 50 kms Southwest of Shillong and is the headquarters of Sohra Civil Sub-division. The total area of this sub-division is about 488 sq. kms (2.18 per cent of the total area of the state) and the population according to 1981 census is 44,500 persons only (i.e., 3.8 per cent of the total population of the state). The rise in the population during 1981-91 is estimated as 31.8 per cent (Census of India 1991).The climate of Cherrapunjee is monsoonal subtropical with distinct wet and dry periods. The year can be divided into a long rainy season (May to October), and a dry winter (November to February) with a short spell of spring during March and April. The annual rainfall at Cherrapunjee ranges between 6689 and 16487 mm with an average of 11276 mm (from 1977 to 1988) (Fig.- 4.1). Cherrapunjee is known as the world's number one inhabited place which receives highest amount of rainfall. The rainfall was as high as 22987 mm in 1961 and 24600 mm in 1974. A major portion (95per cent) of annual rain falls during May to October and rest 5 per cent is received during November to March. The mean monthly maximum and minimum temperatures in the coldest month (January) are 16.2 and 7.7^{0}C and in the hottest month (August) are 23.1 and 18.4^{0}C respectively. The atmosphere is almost saturated with water vapour (relative humidity >90 per cent) during rainy season, but air is relatively dry (relative humidity 60-70 per cent) during the winters. Strong wind (locally known as *Pythaw*) lashes out the area during spring season.

Soil Fertility

The physico-chemical conditions of soil at Cherrapunjee are at sub-optimal level to support healthy plant growth (Table-4.1). The substrata supporting grasslands are stony or gravely. The soil is highly drained and the proportion of clay particles which serve as the repository of plant nutrients is

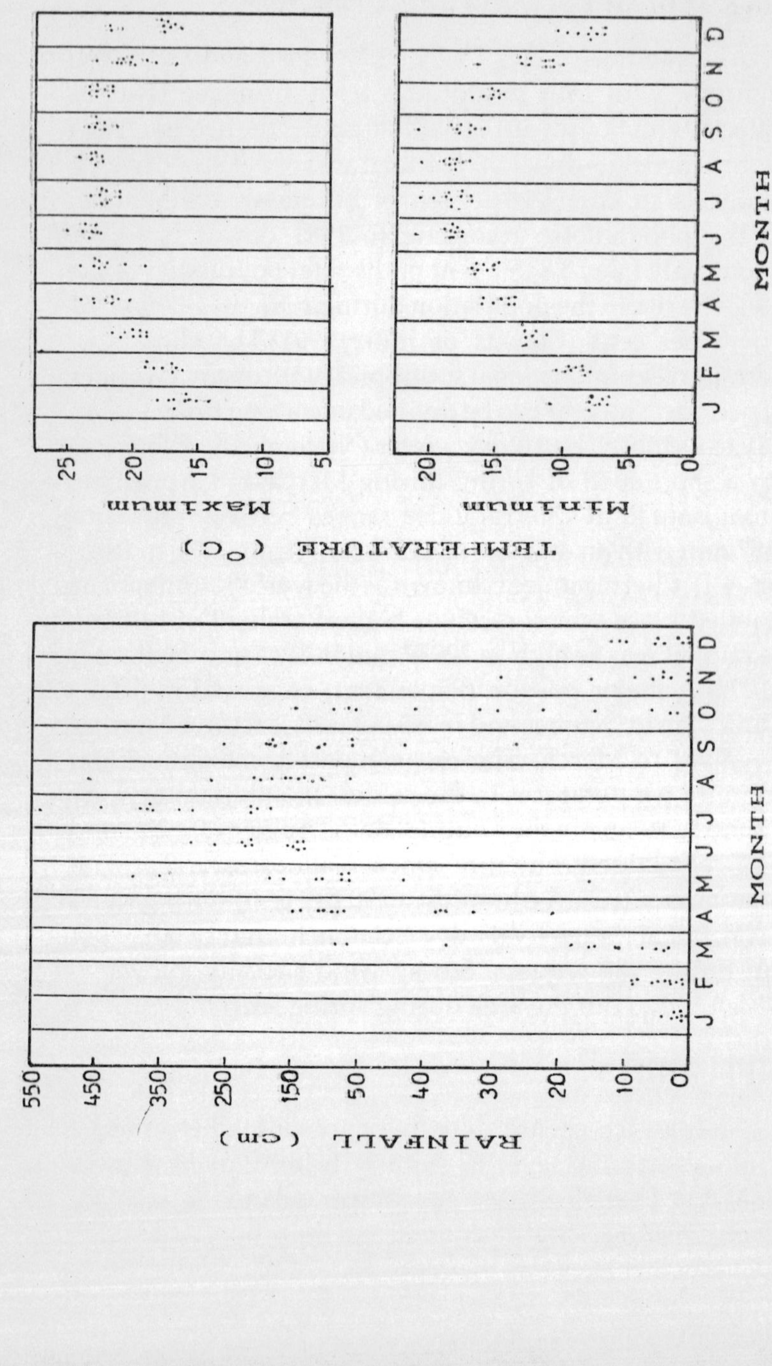

Fig. - 4.1: Variability in Monthly Rainfall and Mean Monthly Minimum and Maximum Temperature during 1977 to 1988.

very low. Nitrogen and Phosphorus, which play a pivotal role in the plant growth, are in extremely low quantities in available nutrient pool in the soils. Approximately 3.8 kg. N/ha and 1.8 kg. P/ha occur in available forms in soil down to 30 cm depth, if the soil is weighted for <2 mm size fraction. A comparison between forest and grassland soils indicates a great deal of similarity between the two. Excessive run off and leaching are the principal factors responsible for this state of soil.

Vegetation

The vegetation of Cherrapunjee can be classified into three categories : (i) broad-leaved evergreen forests, (ii) degraded forest with pine, and (iii) grasslands. The forest is preserved in the form of two small patches, viz., the Sacred Grove and the Reserve Forest. The sacred grove is located around limestone cave in Mausmai village near Cherrapunjee. It occupies about 6 sq. km area and is preserved by the local 'Khasis' due to traditional religious beliefs. The Khasi tribe believes that the sacred grove is the abode of gods/deities and removal of any plant or plant part is against god's will. The reserve forests spread over an area of about 4 sq. kms. This patch of forest is protected by the Village Council. Selective felling of old dying trees is often permitted in this forest. Besides these two protected forests, some broad-leaved and pine trees also grow in isolated pockets on the low lands where the soil and other micro-climatic conditions are conducive for tree growth.

The major species in tree layer of the sacred grove are *Engelhardtia spicata, Syzygium macrocarpa, Echinocarpus dasycarpus, Drimycarpus eacemosus* and *Elaeodendron lanceaefolius*. The major shrub species include *Clerodendron nutans, Phlogocanthus curviflorus* and *Camellia caudata*. The important species which grow in the herb layers along with grasses and ferns are *Boehmeria platyphylla* and *Rubus moluccanus*. Orchids, mosses and climbers are abundant in the forest. The reserve forest is dominated with trees like *Lingustrum robustum, Quercus spicata, Lithocarpus dealbata, Quercus dealbata* and *Wendlandia tinctoria*. The major shrub elements are *Camellia caudata, Eurya japonica* and *Syzygium macrocarpa*. The herb layer is composed of *Strobilanthes coloratus, Carex baccans, Adianthum sp., Panicum maximum, Urena lobata* and saplings of *Gomphostemma parviflora, Camellia caudata, Eurya japonica* and *Clerodendron bracteatum*. Species like *Englehardtia spicata* and *Drimycarpus racemosus, Clerodendron nutans* and *Symplocos glomerata* which are dominant in the sacred grove, occur occassionally in the reserve forest. However, in both the forests evergreen species are dominant. The number of deciduous species is much smaller in the sacred grove compared to the reserve forest. Understory has relatively more evergreen species compared to overstory in both the forests (Table - 4.2).

Broadly speaking, grasslands of Cherrapunjee come under Themeda-Arundinella type (Dabadghao & Shankarnarayan 1973). At local level, five grassland types, namely, *Osbeckia type, Arundinella benghalensis* type, *Ischaemum* type, *Eragrostiella* type and *Arundinella khaseana* type have been recognised by Ram (1986) and Uma Shankar (1991), depending on frequency of burning, soil characteristics and species composition. *Osbeckia* type community generally occurs where the soil is relatively thick and burning is infrequent (once in six years). In this community, dominant species are *Osbeckia crinita, Arundinella benghalensis, Chrysopogon gryllus* and *Fimbristylis thomsonii*. The *A. benghalensis* type community is similar to the *Osbeckia* type, but has *Arundinella benghalensis, Chrysopogon gryllus* and *Carex cruciata* and *Eulalia trispicata* as dominant species.The *Ischaemum* type community developed on sites subjected to burning at two year interval is dominated by *Ischaemum goeblii* and *Fimbristylis complanata*. The *Eragrostiella* type community found on unburned sites with ill weathered, shallow (<30 cms) and nutrient deficient soils is composed of *Eragrostiella leioptera, Borreria articularis* and *Fimbristylis complanata. A. khaseana* type community is developed on hilly coarsed-textured and nutrient impoverished soil. The site is protected from burning and regular livestock grazing for about four years. On this site, *Arundinella khaseana, Chrysopogon gryllus, Eulalia trispicata* and *Eragrostiella leioptera* are important species in the community.

Table-4.3 lists the species recorded in the five grassland communities. None of the community had more than 21 species. All communities showed high dominance and low species diversity. Leguminous species were either noticeably absent or represented by only a few individuals. In all the communities, perennials were the major structural elements and annuals were under represented.

Biomass and Organic Production

Productivity and biomass data are available only for the grassland communities. Despite differences in total biomass accumulation, these communities are markedly similar in allocating greater proportion of organic matter to the belowground parts (Fig.-2b). In this respect, they are more close to the temperate grasslands. Greater allocation of resources to the belowground parts is regarded as an adaptive strategy for survival under stress conditions (Chapin 1980) and for maximizing the root surface area for nutrient uptake (Nye 1977). Most of the belowground biomass (73-80 per cent) in these communities is concentrated in 0-10 cms soil layer, probably due to rocky nature of substrata which do not allow the roots to penetrate to deeper layer.

Fire is one of the most important factors which determine the growth and vigor of plants and control the overall functioning of ecosystem. The *Osbeckia* community which was not subjected to burning for last six years had maximum aboveground biomass followed by *A. khaseana* community which was not burned for four years (Fig.-4.2a). In *A.*

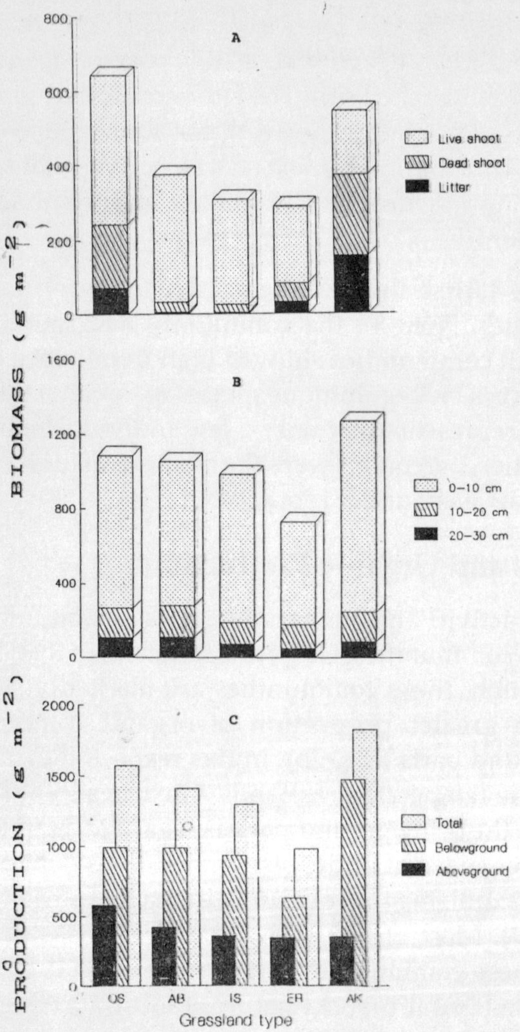

N.B. : OS= Osbeckia type, AB=Arundinella benghalensis type, IS=Ischaemum type, ER=Eragrostiella type, AK=Arundinella khaseana type.

Fig - 4.2 : Average Dry-matter Standing State in Aboveground Compartments (A) and in three Soil Layers (B), and Net Primary Productivity (C) in Five Grassland Communities of Cherrapunjee *benghalensis* community where burning took place just one month prior to the beginning of the study, biomass was substantially low. The *Ischaemum* type community which was exposed to regular burning at two years interval and where last burning was done one month before the start of study had about one-half biomass of the *Osbeckia* type community. The *Osbeckia* type and *A. khaseana* type communities had a fairly good amount of standing dead and litter mass on the ground surface, while in *A. benghalensis* and *Ischaemum* types, the amount of standing dead and litter mass was much lower. Besides fire, soil conditions also influence biomass accumulation in the community. This is clearly evident in case of *Eragrostiella* community which had the lowest amount of biomass even without burning.

Belowground Net Primary Production (BNP) was more than the Aboveground Net Primary Production (ANP) in all the communities (Fig.-4.2c). This was particularly true where *Arundinella khaseana* and *Chrysopogon gryllus* were abundant. The ANP was maximum in *Osbeckia* community which was not burned for six years. ANP declined following fire. Ramakrishnan and Ram (1988) reported that biomass and productivity declined in the subsequent year of burn and it took about two years to reach pre-burn levels. This time lag in the recovery process is ascribed to the loss of nutrients from the system through ash and sediment due to heavy rainfall.

Nutrient Accumulation and Dynamics

Nutrient dynamics, which involves complex and continuous interactions between biotic and abiotic components of the ecosystem, plays a key role in the growth and development of plant community on degraded sites. Nutrient cycling in an ecosystem is primarily controlled by its static and dynamic properties (Burke 1989). Species composition, nutrient content in soil and soil texture etc. are static properties at time scale of decades, whereas dynamic properties include soil available nutrient pools, soil moisture and temperature and population of microbes in the soil which vary widely over shorter time scales, i.e., daily, monthly, seasonally and/or annually and across the altitudinal gradient. Other factors which influence cycling of nutrients in grassland ecosystems include fertilizer application, burning and climatic conditions of the area.

Our understanding of nutrient accumulation and dynamics in degraded ecosystem at Cherrapunjee is based on the data recently collected from *A. khaseana* community by Uma Shankar (1991). The results of the study reveal that nitrogen, phosphorus and cations in the shoots of the dominant grass species are below the dietary requirements of the livestock (Table-4.4). *C. gryllus* showed greater of N, P, Na, and Ca, whereas *A. khanseana* stored more K and Mg. Such a preferential requirement of dominant species for mineral elements could be one of the important reasons for their co-existence on nutrient poor soil of Cherrapunjee. Because when a community is dominated by the one or a few species, there might be a greater nutrient limitation to the individuals growing on less fertile soils, due to continued demand of the same resource (Chapin *et. al.* 1986).

Annual uptake of nutrients and their distribution in above and below ground parts of the community varied from one element to another. A major portion of N, P and Na uptake was reflected in the below ground parts, whereas Ca, Mg and K were largely translocated to the shoot system (Table-4.5). In an annual cycle a major portion of nutrients which are absorbed by the vegetation is returned to the soil through litter and dead root (N 98, P 77, K 49, Na 109, Ca 87 and Mg 65 per cents). In this process of nutrient release belowground parts play a more important role than the shoot system because of their fast turnover rate. Rain water which serves as an important source of N and P input into the system is also responsible for loss of nutrients, especially cations, through run-off and leaching (Table-4.7).

Role of Fire

At Cherrapunjee burning of the dried vegetation during dry winter months (January-February) is an integral component of socio-cultural milieu of the people. The impact of fire on ecosystem is profound and its consequences are dependent on intensity and frequency of fire, proportion of biomass that is burned, the time of monsoon setting and total annual precipitation.

The direct influence of fire on vegetation is the conversion of dried shoot and dead remains on the ground surface into ash and other combustion products. The extent to which organic matter is transformed into ash depends on a number of factors, such as, intensity and duration of fire, fuel load, moisture content in the fuel, weather and topography (Boerner 1982). After fire, the nutrients locked in organic matter may be:

- deposited as ash on the soil surface,
- held up in charred but not completely burned material,
- lost to atmosphere through volatilization.

Nitrogen and sulfure which are volatilized at relatively low temperatures (Ca. 200°C) are readily freed into atmosphere in proportion of the amount of fuel burned (White *et. al.* 1973). Phosphorus and cations, viz., potassium, calcium, sodium and magnesium are generally lost in low quantities because their volatilization temperatures are relatively higher (>500°C) (White *et. al.* 1973). Nonetheless, large amounts of cations may be lost during fire by convection of ash (Harwood and Jackson 1975). At Cherrapunjee, study of soil properties immediately after burning of aboveground vegetation showed an increase in pH and cations and a decrease in carbon and nitrogen contents in the surface soil (Table - 4.6) (Ram and Ramakrishnan 1988). Quick release of nutrients, especially cations, soon after the fire has been reported by other workers as well (Kellman *et. al.* 1985).

Rainfall that follows fire, is of particular importance because it causes heavy loss of nutrients. At Cherrapunjee a major portion of rainwater is lost from the system by way of run off and the loss by percolation though soil profile is relatively less important (Ram and Ramakrishnan 1988). This implies that a major portion of nutrients released after burning of phytomass is lost through run off water and relatively less amount enters into the soil profile (Table-4.7). At many places, dissolved nutrients in soil solution move upward by capillary action during the dry winter period, but this sort of nutrient movement is not possible at Cherrapunjee due to hilly topography and coarse texture of soil.

Thus, combustion of dried phytomass which forms an effective covering on the soil surface and serves as the energy and nutrient source to soil micro-organisms alters the physico-chemical and biological properties of soil and strongly influences the hydrology and nutrient cycling by accelerating losses through various channels. Besides these effects, fire seems to be responsible for bringing about change in the species composition of the community by causing death of those species which propagate mainly through seeds. This might be one of the reasons for poor representation of dicot species including legumes in the Cherrapunjee grasslands.

Strategies for Eco-Restoration

The need for ecorestoration of Cherrapunjee is now widely accepted, and due emphasis is being laid on this task by government as well as non-governmental organisations and agencies. The task of restoring degraded land is not an easy one, because it requires much clearer understanding of soil, plants and climate and their interactions with each other. It also involves understanding of how land is degraded and disturbed and which factors are of critical importance (Bradshaw and Chadwick 1980).

While formulating strategies for ecological restoration

for any area, the first step is to decide the priorities based on the culturable potential of the land in question. The foremost priority must be given to the essential needs of the people inhabiting the area. The necessary requirements of the people who would essentially include the basics of livelihood foodgrains and fruits, fiber, fuelwood and timber, and fodder for livestock. To meet the foodgrain requirements, agricultural use of the land should be preferred.

Areas where sustained agriculture is possible should be marked and brought under cultivation. Where it is not practicable, possibility of plantation forestry could be explored. In plantation, forestry emphasis is most often laid on the fast growing tree species with a view to obtain high economic returns in the form of timber, fibre, fodder or fruits within a short duration; and what is ignored is the carrying capacity of the soil. However, if it is realised that the soil is not able to support healthy growth of plantation forestry, the landscape must be converted into grasslands or it can even be restored for amenity purposes such as recreational lands - playing fields, parks, running tracks etc.

At Cherrapunjee presence of small pockets of crooked forests and growth of grasses, sedges and of members of melastomaceae on exposed surfaces indicate that the micro-level differences in edaphic conditions are very wide. And probably this is the main reason why both forests and grasslands are existing as stable communities under the same climatic conditions. The former representing the climatic climax and the latter exhibiting the bio-edaphic climax community in the area. Since these two types of communities differ markedly from each other in their structure and function, any uniform method of their ecorestoration will not yield desired results. Therefore, it is essential to adopt different strategies for different eco-system. The sacred forest requires protection from all types of human interventions for its growth and as a repository of rich gene-pool. In the grassland areas efforts should be directed mainly to minimize sediment loss and conserve moisture and nutrients in thin soil so that quality and quantity of grassland produce could be gradually improved by encouraging growth of better graminaceous and leguminous species in the community. Some of the strategies which could be taken into consideration for the ecorestoration of Cherrapunjee are as follows :

1. Conservation of Gene-pool and Bio-diversity : The origin and the extinction of species are the two continuously occurring processes in nature. The human interference with nature has, however, accelerated the latter process to such an extent that, if unabated, it would cause a massive loss of gene pool even before we realize the economic importance of several taxa and derive benefits from them. It is, therefore, necessary to preserve them because of the following reasons :

- we don't know at present which species would be of importance and to what extent in future,
- if the gene-pool is once lost, it cannot be retrieved again,
- the posterity must not be deprived of the nature's gift, and
- biological diversity is the basis of all bio-productivity.

The most simple, cost-effective and appropriate effort for the conservation of biodiversity is to provide protection. A number of temple forests (sacred groves) and scattered patches of crooked pocket forests are the potential repositories of biodiversity at Cherrapunjee. Such areas must be protected and preserved. Removal of any plant part should be completely avoided and care must be taken that the litter and detritus is not destroyed by fire.

2. Conservation of Grass Cover : Conservation of the existing grass cover at Cherrapunjee is necessary because :

- the dominant perennial grasses in the community have been found to apportion a major portion of their annual net primary productivity to the belowground parts, which form a stout mat of roots and rootstocks and help bind greater amount of soil,

- the perennial species have the advantage of regenerating through rhizome and don't depend on seeds which may be washed away due to exceptionally high rainfall,
- quick growth of these species in the beginning of rainy season is responsible for the transfer of a major quantity of available nutrients from the soil to phytomass, thus contributing to their conservation against run off and leaching losses,
- dead shoot and litter of the grasses from a protective shield on the ground surface, which prevents slashing of soil particles,
- the dead shoot and root mass also help in soil formation process and create favorable micro-environmental conditions forgreater microbial growth (Fig.-4.3 A and B), and
- the nutrient poor coarse-textured soil is incapable of sustaining healthy tree growth.

3. **Restoration of Degraded Forests :** The idea of forest development by planting trees on open upland sites doesn't seem tenable at Cherrapunjee because of poor quality of soil, and unfavorable micro-environmental conditions. In recently initiated restoration through social forestry plantation at Cherrapunjee what has been ignored is the tenet of succession theory which predicts that late successional species will tend to fail when introduced into a disturbed or open site due to unfavorable water and nutrient relations. During normal succession process, the growth of the vegetation is favoured by the ameliorated site conditions as a consequence of subtle adaptations and interactions within the antecedent vegetation. The ecorestoration of such areas can be best initiated by

Fig. -4.3 A : Soil Profile (foreground) and Pocket Forest Growth amidst Grassland Matrix (background) at Cherrapunjee.

Fig. -4.3 B : Vast Extension of Grassland and Scattered Trees Growing along the Streams.

growing early successional species which are suited to the local conditions. It is, therefore, essential to have a thorough understanding of ecological requirements of the individual species and their subtle adoptions and interactions within a community before selecting them for plantation.

Presence of patches of degraded forests in the area provides clue that these degraded forest sites could be ideal for the forest restoration work, since the site quality is relatively better for the success of plantation forestry. In order to understand the requirements of the tree plantings and their influence on soil conditions, a good deal of research is required through carefully designed experiments under the field conditions.

4. Exclusion of Fire and Control of Grazing : Natural ecosystems exhibit considerable resistance and resilience to periodic disturbance. But persistent or continued disturbance may have pronounced and prolonged ill effects on ecosystem (Odum 1983). Ground fire introduced almost every year during dry winter period in the grasslands at Cherrapunjee is the chief disturbing force causing major nutrient losses from the system. Therefore, there is an urgent need for controlling fire in the area. Grazing is also prevalent despite the poor forage quality. It not only adversely affects grassland communities, but also damages the planted as well as naturally recruited tree seedlings. Prevention of grazing is not possible because of scarcity of fodder in the area, but it could possibly be regulated by adopting scientific methods of pasture management.

5. Increasing Awareness among the Local People : Full support of the local people is a pre-requisite for the success of any ecorestoration work. The people inhabiting the degraded area are the worst sufferers and they fully realize the adverse consequences of degradation of vegetation and soil, even then they are often compelled to cause further damage under pressing need of fuel and fodder. In spite of this, it is imperative to explain them the causes and consequences of land degradation and the underlying principles of ecorestoration. By organising frequent meetings and by using audio-visual aids and other means, ecological awareness could be generated among the local people and they could be actively involved in the ecorestoration activities.

Acknowledgment

The authors gratefully acknowledge the financial assistance from the Ministry of Environment and Forests, Government of India, New Delhi (Grant No. 14/199/84-MAB/RE).

References

Allen, B.C., (1905) (reprint 1980) : *Gazetteer of the Khasi & Jaintia Hills, Garo Hills, Lushai Hills.*, Mittal Publications, Delhi.

Boerner, E.J., (1982): Fire and nutrient cycling in temperate ecosystems. *Bioscience* 32: 187-192.

Bradshaw, A.D. & Chadwick, M.J., (1980): Restoration of land: the ecology and reclamation of derelict and degraded land. *Studies in Ecology* No. 6. Blackwell Scientific Pub. p. 317.

Burke, I.C., (1989): Control of nitrogen mineralisation in Sagebrush steppe landscape. *Ecology* 70: 1115-1126.

CAB (Commonwealth Agricultural Bureau) (1980): The nutrient requirements of ruminant livestock. *Tech. Rev.* CAB, England.

Census of India, (1991): *Provisional Population Totals*, Series-16. Directorate of Census Operation, Meghalaya, Shillong.

Chapin, F.S. III, (1980): The mineral nutrition of wild plants. *Ann. Rev. Ecol. Syst.* 11: 233-260.

Chapin, F.S. III, Vitousek, P.M. & Van Cleve, K., (1986): The nature of nutrient limitation in plant communities. *Am. Nat.* 127: 48-58.

Dabadghao, P.M. & Shankarnarayan, K.A., (1973): *The grass cover of India.* ICAR, New Delhi.

Harwood, C.E. & Jackson, W.D., (1975): Atmospheric losses of four plant nutrients during a forest fire. *Aust. For.* 38: 92-99.

Kellman, M. Miyanishi, K. & Kiebert, P., (1985): Nutrient retention by savana ecosystem, II: Retention after fire. *J. Ecol.* 73:953-962.

Kemp, A., (1960): Hypomagnesaimia in milking cows: the response of serum magnesium to alterations in herbage composition resulting from potash and nitrogen dressing to pasture. *Neth. J. Agric. Sci.* 8: 281-304.

Khiewtam, R.S., (1986): Ecosystem function of protected forests of Cherrapunjee and adjoining areas. *Ph.D. thesis,* North-Eastern Hill University, Shillong. p. 182.

National Research Council, (1971): Nutrient requirements of diry cattle. 5th edn., *Nat. Acad. of Sciences,* Washington, D.C.

Nye, P.H., (1977): The rate limiting step in plant nutrient absorption from Soil. *Soil Sci. Soc. Am. Proc.* 24: 33-35.

Odum, E.P., (1983): *Basic Eclogy.* Philadelphia, Saunders College Publishing. p. 613.

Ram, S.C., (1986): Ecosystem structure and function of seral communities of degraded environment at Cherrapunjee and adjoining areas. *Ph.D thesis,* NEHU, Shillong. p. 218.

Ram, S.C. & Ramakrishnan, P.S., (1988): Hydrology and soil fertility of degraded grasslands at Cherrapunjee in North-Eastern India. *Environ. Conserv.* 15: 29-35.

Ramakrishnan, P.S. & Ram, S.C., (1988): Vegetation, biomass and productivity of seral grasslands of Cherrapunjee in North-Eastern India. *Vegetatio* 84: 47-53.

Uma Shankar, (1991): Nutrient cycling in degraded ecosystems (grasslands) of Meghalaya. *Ph.D. thesis,* NEHU, Shillong. p. 187.

White, E.M., Thompson, W.W. & Gartner, F.R., (1973): Heat effects on nutrient release from soils under Ponderosa pine. *J. Range Manage.* 26: 22-24.

Wordon, A.N., Sellers, K.C. & Tribs, D.E. (eds.), (1963): *Animal Health, Production and Pasture.* Longman, London.

Table-4.1 : Physical and Chemical Characterstics of the Soil (down to 30 cm. depth) of the Sacred Grove, Reserve Forest and Five Grassland Communities at Cherrapunjee.

Character	Sacred Grove **	Reserve Forest **	Grassland Communities*				
			Osbeckia	A.benghlensis	Ischaemum	Eragrostiella	A #. khaseana
(a) Soil Substratum Composition (%)							
Stone (>5 mm)	-	-	-	-	-	-	40.0
Gravel (5 to 2 mm)	-	-	-	-	-	-	28.1
Mineral Particles (<2 mm)	-	-	-	-	-	-	23.9
(b) Texture (% of Mineral Particles)							
Sand	-	-	-	-	-	-	80.7
Silt	-	-	-	-	-	-	10.3
Clay	-	-	-	-	-	-	9.0
Water Holding Capacity (%)	-	-	-	-	-	-	41.9
pH	4.4	4.2	5.5	5.5	5.4	5.2	5.5
Organic Carbon (%)	2.4	2.3	2.6	2.4	1.8	1.0	1.5
Total Nitrogen (%)	0.30	0.26	0.29	0.28	0.23	0.14	0.15
C/N Ratio	8.8	9.1	8.9	8.4	8.0	7.1	9.7
NH_4-N (mg $100g^{-1}$)	-	-	-	-	-	-	0.12
NO_3-N (mg $100g^{-1}$)	0.16	0.15	0.29	0.28	0.23	0.20	0.41
PO_4-P (mg $100g^{-1}$)	0.13	0.11	0.18	0.16	0.13	0.12	0.56
CEC (mg $100g^{-1}$)	-	-	-	-	-	-	8.1
Exchangeable Potassium(mg $100g^{-1}$)	0.27	0.26	0.18	0.17	0.17	0.14	0.07
Exchangeable Sodium (mg $100g^{-1}$)	-	-	-	-	-	-	0.06
Exchangeable Calcium(mg $100g^{-1}$)	1.26	0.54	1.38	1.28	1.10	0.73	1.13
Exchangeable Magnesium (mg $100g^{-1}$)	4.24	1.21	1.93	1.69	1.51	1.31	0.21

Sources : ** after Khiewtam (1986); * after Ram & Ramakrishnan (1988); # after Uma Shankar (1991)

Table-4.2 : Important Species of Sacred Grove and Reserve Forest at Cherrapunjee.

Species	Sacred Grove	Reserve Forest
Tree layer		
Ardisia florinunda Wall	-	+
Camellia caudata Wall	+	+
Castanopsis kurzii (Hance.) S.N. Biswas	+	+
Cinnamomum tamala Fr. Nees	+	+
Clerodendron bracteatum Wall	-	+
Clerodendron nutans Wall	+	+
Daphne cannabina Wall	+	+
Debregeasia wallichiana Wedd	-	+
Drimycarpus racemosus Hk.f.	+	+
Echinocarpus dasycarpus Benth	+	-
Elaeocarpus lanceaefolius Roxb	+	-
Englhardtia spicata Bl.	+	+
Eurya japonica Thunb	-	+
Glochidion multiloculare Muell	+	+
Glochidion thomsoni Hk.f.	+	-
Lingustrum robustum Bl.	+	+
Lindera melastomacea Benth	-	+
Lithocarpus dealbata (Miq.) Reh.	+	+
Ltsea elongata Benth. & Hk.f.	+	+
Myrica esculenta Buch. -Ham. ex D.Don	-	+
Phlogocanthus curviflorus Nees	+	-
Quercus dealbata Hk.f. & T.	+	+
Quercus spicata Sm.	+	+
Strobilanthes coloratus T. Anders.	+	-
Symplocos glomerata King ex Gamble	+	-
Symplocos racemosa Roxb.	-	+
Syzygium macrocarpa Roxb.	+	+
Syzygium tetragonum Kurz.	+	+
Wendlandia tinctoria DC.	-	+

Contd. ...

Species	Sacred Grove	Reserve Forest
Herb layer		
Arisaema echinatum Schott.	+	-
Boehmeria platyphylla D.Don.	+	-
Carex baccans Nees	-	+
Fragaria indica Andr.	-	+
Hedyotis scandens Roxb.	+	-
Inula cappa DC.	-	+
Osbechia crinita Benth.	-	+
Panicum maximum Wall.	+	+
Paspalum orbiculare Forst.	+	-
Poa khasiana Stapf..	+	+
Pogostemon sp.	-	+
Robus moluccanus L.	+	-
Selaginella sp.	+	+
Setaria glauca Beauv.	-	+
Utrica parviflora Roxb.	+	-
Viburnum foetidum Wall.	-	+
Ferns		
Adiantum sp.	+	+
Microsorium sp.	+	-
Pteris sp.	+	-
Thelypteris sp.	+	-

Source : Khiewtam (1986).

Table-4.3 : Floristic Composition of Five Grassland Communi ties at Cherrapunjee.

Species	Grassland Communities				
	Osbeckia	A.bengh-lensis	Isch-aemum	Eragro-stiella	A. khaseana
Grasses					
Arundinella benghalensis (Spreng.) Druce	+	+	+	+	+
A. khaseana Nees	-	+	+	+	+
A. nepalensis Trin	+	+	+	-	-
Axonopus compressus (SW.) P. Beauv.	-	-	+	-	-
Capillipedium assimile (Steud.) A. Camus	+	-	+	-	-
Chrysopogon gryllus (L.) Trin	+	+	+	+	+
Dimeria fuscescens Trin.	-	-	+	+	-
Eragrostiella Teioptera (Stapf.)Bor	-	-	+	+	+
Eulalia trispicata (Schult.) Henr.	+	+	+	-	+
Fulalia quadrinervis (Hack.) O. ktze	+	+	-	-	-
Hetropogon contortus P. Beauv	+	-	-	-	-
Ischaemum goeblii Hack.	-	+	+	+	-
Paspalum orbiculare Forst.	+	+	+	+	+
Pogostemon quricularis (L.)Hoask	+	+	+	+	-
Setaria glauca P. Beauv.	+	+	+	-	-
Sedge					
Carex cruciata Wahenb.	+	+	+	+	-
Fimbristylis complanata L.	-	+	+	+	-
Fimbristylis dichotoma (L.) Vahl	-	-	-	-	+

Species	Grassland Community				
	Osbeckia	A.bengh-lensis	Isch-aemum	Eragro-stiella	A. khaseana
Fimbristylis thomsonii Beeck	+	-	+	+	-
Legume					
Eriosema chinense Vogel.	-	-	-	-	+
Forb					
Anotis oxyphylla Hk.f.	-	-	-	+	-
Anotis wightiana Hk.f.	-	-	-	-	+
Borreria articularis (L.f.) N.F. William	-	-	-	+	-
Centella asiatica (L.) Urb.	-	-	-	-	+
Cyanotis vaga (Lour.)	-	-	-	-	+
Schult Eriocaulon cristatum Mart.	-	-	-	-	+
Gentiana quadrifaria Clark non Bl.	-	+	+	+	+
Leucas ciliata Benth.	+	-	-	-	-
Melastoma malabathricum L.	+	+	-	-	-
Osbeckia capitata Benth.	+	+	+	+	+
Osbeckia crinita Benth.	+	+	+	+	-
Phyllanthus simplex Roxb.	+	+	+	+	-
Plactranthus japonicus (Burn f.) Koidz.	+	+	+	-	-
Polygonum bistorta L.	-	-	-	-	+
Total Species (34)	18	18	21	17	15

Source : ** after Ramakrishnan & Ram (1988); and Uma Shankar (1991).

Table-4.4: Nutrient Concentration (mg g^{-1}) in Shoots of Dominant Grasses in A. Khaseana Community at Cherrapunjee, and Dietary Requirement of Six Mineral Elements.

Species	Nitrogen	Phosphorus	Potassium	Sodium	Calcium	Magnesium
A.khaseana	4.80+0.24	0.72+0.04	5.72+0.60	0.24+0.02	1.02+0.04	3.39+0.23
Chrysopogon gryllus	5.97+0.30	0.79+0.05	3.80+0.38	0.25+0.01	1.66+0.10	1.00+0.17
Eulalia trispicata	8.19+0.46	0.80+0.05	3.06+0.30	0.27+0.02	2.66+0.18	1.31+0.10
Eragrostie leioptera	6.65+0.31	0.48+0.03	2.43+0.30	0.18+0.01	1.85+0.14	0.68+0.08
Other Species	8.47+0.55	1.01+0.06	3.12+0.22	0.24+0.01	3.64+0.44	2.19+0.35
Dietatry requirements	11.7-18.0[a]	1.5-3.0[b]	5.0-8.0[c]	0.9-2.0[a]	3.5[c]	2.0[d]

Source : a - CAB (1980); b - Wordon *et al.* (1963); c - National Research Council (1971); d - Kemp (1960)

Table - 4.5 : Annual Uptake and Release (Kg /ha) and Recycling Coefficient (release/uptake) of Six Mineral Elements in A. Khaseana Community at Cherrapunjee.

	N	P	K	Na	Ca	Mg
Uptake						
Aboveground	24.6	3.7	25.7	1.1	6.6	10.0
Belowground	112.7	6.8	25.4	4.4	2.1	8.2
Total	137.3	10.5	51.1	5.5	8.7	18.2
Release						
Through litter	18.2	0.9	3.6	0.5	3.8	1.7
Through root	116.1	7.1	21.6	5.5	3.8	10.0
Total	134.3	8.0	25.2	6.0	7.6	11.7
Recycling coefficient						
Above ground	0.74	0.24	0.14	0.45	0.57	0.17
Below ground	1.03	1.05	0.85	1.26	1.81	1.22
Total	0.98	0.77	0.49	1.09	0.87	0.65

Table-4.6 : Changes in the Chemical Properties of Surface Soil (0-5 cm.) after Fire in Two Grassland Communities at Cherrapunjee.

Character	A. benghalensis		Ischaemum	
	Before fire	After fire	Before fire	After fire
pH	5.80	6.20	5.60	5.90
Organic Carbon (%)	2.35	2.13	1.84	1.78
Nitrogen (%)	0.28	0.25	0.23	0.22
C/N ratio	8.40	8.50	8.00	8.10
NO_3 -N (mg $100g^{-1}$)	0.28	0.29	0.23	0.24
PO_4 -P (mg $100g^{-1}$)	0.16	0.18	0.13	0.14
K (mg $100g^{-1}$)	0.17	0.47	0.17	0.35
Ca (mg $100g^{-1}$)	1.28	2.10	1.10	1.34
Mg (mg $100g^{-1}$)	1.69	2.35	1.51	1.75

Source : Ram & Ramakrishnan (1988).

Table -4. 7 : Annaul Input of Nutrients (gm^{-2}) through Rainfall and their Losses through Run-off (R) And Percolation (P) in Three Grassland Communities at Cherrapunjee.

Character	Input through rain water	Output					
		Osbeckia		Ischaemum		Eragrostiella	
		R	P	R	P	R	P
Water	1038	719.70	136.40	818.20	126.60	856.50	130.70
NO_3-N	4.67	1.86	0.70	4.28	0.58	1.25	0.36
PO_4-P	4.15	0.18	0.03	0.22	0.03	0.17	0.03
K	24.92	4.14	0.83	7.53	0.44	3.38	0.30
Ca	58.14	11.40	1.73	8.41	1.63	7.44	1.42
Mg	22.87	3.92	0.44	3.76	0.64	2.46	0.26

Source : Ram & Ramakrishnan (1988).

Section-B

Problems of Hill Farming

5

Livestock Farming in the Areas of Shifting Cultivation *

Arun Verma
Head of the Animal Sciences,
ICAR Research Complex for North-Eastern Hill Region
Shillong.

In the past, every effort has failed in improving the economy of Jhumia (shifting cultivators of North-Eastern India) inspite of the fact that funds were allotted as per the demand. Our political systems are such that funds are made available may be at wrong time.

The one key factor is our poor capability to utilize resources sufficiently whether they are financial resources, manpower resources or agro-climatic resources. In the present paper, complex of technological, sociological, environmental and economic problems are tried to solve through efficient use of sociological, technological and environmental resources.

Dominant role of Animal Husbandry in Indian Economy has been realised after 42 years of our independence when

---

* Paper presented on National Environmental Awareness Campaign, Workshop on Environmental Problems of East Khasi Hills District of Meghalaya, 28th November 1989.

Planning Commission has announced that in the total GNP of Rs. 190,000 crores livestock contributes 31000 crores which is 15.5 per cent of the total. Some how livestock has still been allotted secondary importance every where. In case of Jhumia, this role has been very well realised. Mithun, Pig, Goat, Cattle and Red Jungle Fowl are close associate to Jhumia. They used to live in perfect harmony with the nature till the early half of this century. The major change has occurred in the last 40 years after the partition of the country. This was the time when major sociological changes in this part of the country took place. And perhaps this was the time when technological advancement should have been brought in. Due to this gap, the livestock which used to be the pride possession has become burden and liability. The major degradation in health and productivity of animals have occurred. This has affected the economy of Jhumia very adversely.

The present paper deals with technologies related with improving livestock productivity in the areas where Jhumming (shifting cultivation) is the major land use system. Jhumming in humid tropics of South-East Asia and Savanah of Africa differ significantly. The present paper deals with the Jhumming of high rainfall areas. In the areas of Jhumming livestock can be subsidiary source of income for those farmers who do not prefer a quantum jump from the traditional agriculture to intensive livestock production. This system is known as subsidiary source of income through livestock. The other system is integrated watershed based land use system for intensive livestock production. The third system is where livestock are dependent on pasture only.

Livestock as Subsidiary Source of Income

A multidisciplinary research project was initiated on studying the potentiality of livestock production as subsidiary source of income for farmers practising jhumming for improvement of jhumming.

Shifting cultivation (W1) was compared with agri-horticulture on bench and half moon terraces (W2), agri-culture bench terraces (W3) and agriculture on puertorican terraces (W4). Livestock was subsidiary source of income in all the above systems. It was observed that in terms of capability of supporting animal unit as subsidiary source of income on the basis of available by-products, forages from the waste and unutilized surfaces, the landuse systems W2, W3 and W4 are more efficient than that of shifting cultivation (W1).

Total run-off in four systems were 114.47, 91.00 and 156.00 mm. Soil losses were 41, 3, 2, and 17 t/ha respectively. Average rainfall was 1628 mm. All these data are average of 3 years except data on W1 which is as per the two Jhumming (shifting) cycle. It is thus evident that agri-horticultural system (W3) are within the permissible limit of soil loss.

It may be inferred that agri-horticultural system (W2) helps in conservation of soil and has the potentiality of providing balanced ration to livestock. Livestock in return provide income through their produce including manure for raising fertility of the land.

Intensive Watershed Based Livestock Production System

For watershed development following expenditure are involved in one hectare area. On every fourth and fifth year 50 per cent area is given rest (Table- 5.1).

This landuse is very much congenial from soil conservation point of view. No run-off was recorded in the last 4 years. In fodder production policy, annual fodder complement perennial fodders in the first year and from the second year perennials become self sufficient in covering 100 per cent canopy before the erosive rain comes. This landuse system is economically sound as it provides highly palatable greens in Rs. 80 per tonne basis, whereas only harvesting and transport cost of local fodder is more than Rs. 300 per tonne.

Table - 5.1 : Watershed Development in One Hectare

Items	Amount
Contour bund	100 mandays
Earth working	20 mandays
Sowing grass	29 mandays
Sowing lagume	25 mandays
Manuring	10 mandays
Fertilizer application	5 mandays
Weeding	20 mandays
Total	209 mandays
Total labour	Rs. 3135.00
Fertilizer	Rs. 800.00
Seeds/Root slips	Rs. 250.00
Total	Rs. 4185.00

Criteria of Selection of Livestock Units

For the present no systematic approach is followed among the tribals. A study was undertaken to asses the best livestock enterprise which can provide substantial income to the farming family. Besides this, six more points need consideration.

1. The unit should suit to the agro-climatic condition of the region.
2. The unit should not demand more than average family labour, i.e. 500 mandays per year.
3. Unit should provide daily as well as lumpsum income to the farmer.
4. The unit of multi-units should be complementary among each other.

5. The unit should be socio-economically acceptable to the farmer.

6. The unit should generate such products which have easy and acceptable market nearby.

Cattle (beef and milk), pig (pork), poultry (egg and meat) and goat are most common animals of the region. Since goat does not suits to pasture systems, they are replaced by rabbits (fur, wool and meat).

Criterion of Selection

It has been earlier assessed that a farming family of three adult and two minor units can spare 500 mandays per year in any enterprise or professional activity to meet his livelihood.

Productivity of the watershed and mandays available for managing the enterprise were the main criterion for the quantitative assessment of the enterprises. Although in watershed development first year is very crucial when whole watershed is to be developed, but if it is planned in phased manner one can easily spare on mandays per day for the livestock unit.

Economics of Livestock Units

Cow unit is the base of livestock production economics, this provides a daily income of Rs. 12.03 from each cow. Pig and rabbit units are next to these. Economics of all these livestock units may be assessed consulting Tables - 5.2, 5.3 and 5.4. On the basis of above assessment, 8 livestock unit combinations can be made and economics of each can be assessed. But before this each livestock unit would be examined separately (Table - 5.5).

Table - 5.2: Cost that Vary of One Animal of Each Enterprise.

Parameters	Cow		Pork		Beef		Rabbit	
	Rs.	Per cent	Rs.	Per cent	Rs.	Per cent	Rs.	Per cent
Labour	7.5	36	1.20	29	2.50	28	0.096	34
Green	5.8	28	0.92	22	4.64	61	0.058	21
Concentrate	7.5	36	2.00	49	1.88	25	0.125	45

Table - 5.3 : Farm Benefit from Products and By products of Each Animal.

Parameters	Dairy Milk		300 Days Pork		547 Days Beef		90 Days Rabbit Meat	
	Qty Kg.	Price Rs.	Qty Kg.	Price Rs.	Qty Kg.	Price Rs.	Qty Kg.	Price Rs.
Main Product	10	30.00	0.20	6.00	0.30	7.50	0.05	0.175
Manure	20	2.80	3.00	0.39	0.20	2.80	0.20	0.030
By-Product	Skin	0.05	Bristle	0.05	Skin	0.05	Fur	0.16

Table -5. 4 : Marginal Rate of Return from Livestock Unit of Each Enterprise.

(fig. in Rs.)

Farm Economics	Dairy	Pork	Beef	Rabbit
Cost That Vary	20.80	27.75	11.28	27.9
Farm Benefit	32.85	40.25	12.31	36.5
Net Benefit	12.05	14.50	1.03	8.6
Marginal Rate of Return (in %)	57.93	56.31	9.23	30.8

Table - 5.5: Net Gain for Farmers Every Year.

(fig. in Rs.)

Overall Profit	Cow	Pig	Beef	Rabbit
Germplasm	600.00	1875.00	200.00	1000.00
House	1200.00	1200.00	1200.00	1200.00
Gross Income	4398.00	5292.00	375.95	3139.00
Net Gain	2598.30	2217.50	(-)1024.05	939.00

Livestock Dependent on Pasture

Cattle, sheep, goat, mithun and yak are graziers cum browsers. The pasture vegetation provides low protein (3.4 to 4.6 per cent), high fiber (28.7 to 41.8 per cent) situation for livestock. The species bread and individual which has the capacity of high dry matter intake should survive better over the other types.

Planning for Pasture Lands

1. Classification of hill slope suited to specific pasture species.
2. Treatment of hill slope as per the need.
3. Allow grazing if sufficient fodder is there.
4. Stocking density as per the carrying capacity.

Pasture Type

Native, cultivated and sylvipasture may be developed as per the need. Conservation base of contour bund and half moon terrace may be developed for pasture and tree species.

A study on pasture development was conducted in ICAR Research Complex NEHR, Shillong. Salient features of the study are as follows :

The study is described in three parts. In the first, problems of land and area were studied. Soil erosion in cropping was 32.9 to 83.8 t/ha as compared to 0.04 to 0.05 t/ha in forest eco-system. Stocking density of livestock is more than the carrying capacity of pastures. Native pastures are low in phosphorus content (0.073 ± 0.02).

In the second part of the study, cultivated, improved native and sylvipasture were developed in Vth and VIth categories of land. Cultivated pastures had production value of 7.19 ± 1.44 t/ha/yr. dry matter as compared to 2.47±0.02 t/ha/yt of native pastures.

In the third part of the study, affect of grazing on steep hill slopes by goats were studied. At a carrying capacity of 25 goats in well developed pasture soil erosion rate was 0.19 to 0.7 t/ha/yr. Goat grazing did not adversely affect representation of stylo-based pastures.

6

Jhum Cultivation in Meghalaya

B. K. Tiwari
Centre for Eco-Development,
North -Eastern Hill University, Shillong
and
C.B. Pandey
Regional Centre , N.A.E.B.
North-Eastern Hill university, Shillong

Socio-Economic Status of Jhumias in Meghalaya

Shifting cultivation is widely practised in the hill areas of the North-east, Southern Orissa and Northern Andhra Pradesh. It is the main stay of local tribals there. It is also prevalent in isolated pockets in Madhya Pradesh, Maharashtra, Bihar, Kerala and Tamil Nadu. It is known generally as *Jhum* in Meghalaya but as *Tekeonglu* in Nagaland, *Adimdik* in Arunachal Pradesh and *Heckusisomoms* in Tripura; *Bagad, Dqwar, Dipa* and *Dahis* in Madhya Pradesh; *Kumri* or *Kumari* in Western ghats and *Kurai* and *Khalli* in Bihar. During 1960-61 around 618 thousand hectares land was under shifting cultivation in India which increased to 1326 thousand hectare in 1974-75. Meghalaya plateau covers an area of 22540 sq. kms. comprised of different hilly terrains with narrow valleys in between and strips of plain lands adjacent to Bangladesh in the South and West. In the North, it borderes with the great Brahmaputra valley of Assam and, in the East, the detached Mikir Hills.

Physiographically, the Meghalaya plateau is remarkable due to its hilly desected irregular terrain in the Western faces in contrast to the regular and steep fall of the Southern face. Climate is very fluctuating due to seasonal winds. Soil is prominent by red soil, sandy gravel and shale because of heavy rainfall and other geological underlyings.

The economy of the plateau is almost entirely dependent on agriculture, although 85 per cent of the state land surface has been classified as barren and unsuitable for agriculture. Maximum area (760 sq. kms) is under jhum cultivation in Meghalaya and around 20 per cent of its total population is involved in practising jhum cultivation. Jhum is mostly confined to rural area, however, only 0.1 per cent is practised in urban areas also.

Land Use

Only 37 per cent of the total geographical area is under forest. Around 7 to 8 per cent of total geographical area is under the category of net sown area, of which only 1.3 to 1.4 per cent has been recorded as sown more than once. 14 per cent is under non-agricultural uses (Table-6.1). There are in all 5100 villages in Meghalaya, out of which 418 are in Jaintia Hills District, 1227 in East Khasi Hills, 743 in West Khasi Hills, 677 in East Garo Hills and 1383 in West Garo Hills. 61 per cent of the total villages in East garo Hills and 68 per cent in West Garo Hills are involved in jhum cultivation.

A total of 28352 households in Jaintia Hills, 100313 in East Khasi Hills, 29167 in West Khasi Hills, 26406 in East Garo Hills and 71697 in West Garo Hills are practising jhum cultivation (Table- 6.2). Maximum jhum cultivation occurs in West and East Garo Hills.

Jhum Land and Its Ownership

The land ownership is an important factor determining the landuse pattern under shifting cultivation. There are two category of jhum land i.e. owned, and rented. Maximum shifting cultivation lands are generally owned by the village communities. The individual family is assigned a piece of land for cultivation during the season. Thus, he has only right to use it. Maximum 76 to 96 per cent jhumias in different districts of Meghalaya are engaged in owned jhum land and a very little 1.5 to 4.2 in rented and 1.5 to 19.2 per cent in both owned and rented jhum land (Table-6.3). Same way maximum 79.8 to 92.9 per cent population are involved in jhumming on owned land.

Population Involved in Jhum Cultivation by Category of Jhum Land

Average size of jhum land per house hold ranges from 0.4 to 1.3 hectare and the area per jhum plot is 0.3 to 1.1 hectare. However, the land size under permanent cultivation is grater compared to that of jhum land. Of the total population, 9.5 per cent in Jaintia Hills, 6.9 per cent in East Khasi hills, 11.9 per cent in West Khasi Hills, 34.1 per cent in East Garo Hills and 38.3 per cent in West Garo Hills are engaged in jhum cultivation (Table - 6.4). In Laskein and Khliehriat C.D. Blocks in Jaintia Hills, Bhoi area and Nongpoh Blocks in East Khasi Hills, Songsak, Samanda and Dambo Rongjeng in East Garo Hills, and Rongrara, Dambuk, Chokpot, Dalu and Rongram in West Garo Hills, maximum people are involved in jhum cultivation. Maximum jhumming in these areas is practised in owned land, however, in Mawsynram Block of East Khasi Hills, jhumming occurs only in rented or owned land.

Table - 6.1: Landuse Pattern in Meghalaya.

Categories	1973-74	1974-75	1975-76	1976-77	1977-78	1978-79
Geographical Area (in 0000ha)	224.89	224.89	224.89	224.89	224.89	224.89
1. Forest	36.59	36.59	36.60	36.59	36.17	36.13
2. Not Available for Cultivation	13.75	13.75	13.75	13.75	13.92	13.99
3. Other Cultivated Land Excluding Fallow Land	27.29	27.29	27.28	27.28	27.66	27.55
4. Fallow Land	14.66	14.65	14.64	14.64	14.33	13.87
5. Net Area Sown	7.72	7.72	7.73	7.65	7.93	8.57
6. Area Sown More than Once (%)	1.29	1.29	1.29	1.31	1.36	1.36
7. Total Cropped area(%)	9.00	9.02	9.03	9.05	9.29	9.93

Source : *Statistics of Jhum Cultivation in Meghalaya,* The Directorate of Economics & Statistics and Evaluation, Meghalaya, 1988.

Table - 6.2 : Number of Households Involved in Jhum Cultivation.

Sl. No.	District C.D. Block	Total No. of House Holds	% of Total H.H.invol-ved in Jhum	Percent Owned	Jhumia Rented	Household Owned & Rented
(a)	**Jaintia Hills**	**28352**	**11.47**	**89.45**	**4.15**	**6.40**
1.	Thadlaskein	8407	4.80	73.20	3.47	23.33
2.	Laskein	8349	13.57	89.05	6.71	4.24
3.	Khliehriat	7768	16.55	99.77	-	0.23
4.	Amlarem	3828	11.21	74.83	10.49	14.68
(b)	**E. Khasi Hills**	**100313**	**7.01**	**76.84**	**4.01**	**19.15**
1.	Nongpoh	7973	17.77	70.36	0.64	29.00
2.	Bhoi-Area	11912	26.05	90.11	2.13	7.76
3.	Mawryng kneng	4988	2.63	100.00	-	-
4.	Mylliem	44537	1.98	71.77	7.82	20.41
5.	Mawphlang	5358	3.06	82.92	8.54	8.54
6.	Pynursla	7262	8.59	65.54	10.42	24.04
7.	Shella	7688	6.50	34.80	2.60	62.60
8.	Mawsynram	6819	0.23	-	43.75	56.25
9.	Mawkynrew	3776	5.08	65.11	20.31	14.58
(c)	**W.Khasi Hills**	**29167**	**12.44**	**77.13**	**1.49**	**21.38**
1.	Mairang	7923	2.21	30.28	0.05	2.73
2.	Mawkyrwat	8634	5.72	93.32	6.07	0.61
3.	Nongstoin	7598	14.96	69.66	1.05	29.29
4.	Mawshynrut	5012	34.42	86.55	0.46	12.99
(d)	**E. Garo Hills**	**26406**	**33.84**	**96.25**	**2.27**	**1.48**
1.	Dambo Ronggeng	8253	22.99	89.46	4.95	5.59
2.	Songsak	5160	67.67	99.66	0.08	0.26
3.	Resu Belpara (east)	8968	17.18	92.21	6.88	0.91
4.	Samanda	4025	49.81	99.85	-	0.15
(e)	**W.Garo Hills**	**71697**	**40.28**	**93.00**	**3.84**	**3.16**
1.	Resu Belpara (west)	3496	28.63	99.00	0.40	0.06
2.	Dadenggiri	9765	40.19	97.40	0.13	2.47
3.	Selsella	10776	24.32	96.98	1.22	1.80
4.	Rongram	12522	47.71	97.99	0.45	0.66
5.	Betasing	8229	29.63	83.47	6.23	10.30
6.	Zikzak	7961	37.47	95.54	0.54	3.92
7.	Dalu	7042	50.87	97.57	0.35	0.88
8.	Chokpot	5760	46.94	80.51	18.05	1.44
9.	Dambuk Aga	3857	59.06	75.29	15.36	9.35
10.	Rongara	2309	61.84	96.50	0.63	2.67
Meghalaya		**255935**	**20.21**	**90.03**	**3.44**	**6.53**

Source : As in Table - 6.1.

Table- 6.3: Classification of Jhum Land in 1981.

State/District	Type of Jhum Land	Jhumia No.	%	Persons engaged	Per cent to total
	Owned	**46565**	**93.7**	**231557**	**90.12**
Meghalaya	**Rented**	**1787**	**3.6**	**8370**	**3.26**
	Owned & Rented	**3374**	**6.8**	**17020**	**6.62**
	Owned	2908	89.5	13544	90.89
Jaintia Hills	Rented	135	4.2	634	4.25
	Owned & Rented	208	6.4	724	4.86
	Owned	5401	76.8	28356	79.79
East Khasi hills	Rented	282	4.0	1313	3.69
	Owned & Rented	1346	19.2	5871	16.52
	Owned	2799	77.1	15287	79.59
West Khasi Hills	Rented	54	1.5	363	1.89
	Owned & Rented	776	21.4	3557	18.52
	Owned	8600	93.0	42699	91.80
East Garo Hills	Rented	203	2.3	1110	2.39
	Owned & Rented	132	1.5	2703	5.81
	Owned	26857	93.0	131671	92.86
West Garo Hills	Rented	1108	3.8	5950	4.20
	Owned & Rented	912	3.2	4165	2.94

Source : As in Table- 6.1.

Table -6.4 : Population Involved in Jhum Cultivation by Category of Jhum Land.

District/ C.D. Block	Population	People engaged in jhum cultivation (% to total population)	Percentage of populatioin engaged in jhum cultivation		
			Owned	Rented	Owned & Rented
Jaintia Hills	**156402**	**9.53**	**90.89**	**4.25**	**4.86**
1. Thadlaskein	44709	2.79	74.06	3.28	22.66
2. Laskein	46853	12.57	87.79	7.66	4.55
3. Khliehriat	43075	15.06	99.92	-	0.08
4. Amlarem	21765	5.86	75.70	11.13	13.17
East Khasi Hills	**511414**	**6.94**	**79.79**	**3.69**	**16.52**
1. Nongpoh	39555	18.19	-	-	-
2. Bhoi-Area	60378	25.88	92.37	1.61	6.02
3. Mawryngkneng	27132	2.30	100.00	-	-
4. Mylliem	229792	2.45	78.63	5.37	16.00
5. Mawphlang	29455	3.29	83.87	8.99	7.64
6. Pynursla	34001	7.96	63.33	12.59	24.08
7. Shella Bholaganj	36549	4.49	36.50	3.60	59.90
8. Mawsynram	34137	0.28	-	38.30	61.70
9. Mawkynrew	20415	5.11	68.93	17.55	13.52
West Khasi Hills	**161576**	**11.89**	**79.59**	**1.89**	**18.52**
1. Mairang	45023	2.75	21.05	1.77	77.18
2. Mawkyrwat	46206	5.93	91.50	7.77	0.73
3. Nongstoin	42185	14.83	74.61	1.41	23.98
4. Mawshynrut	28162	31.85	87.52	0.04	12.04
East Garo Hills	**136550**	**34.06**	**91.80**	**2.39**	**5.81**
1. Dambo Ronggeng	44329	27.42	74.03	4.63	21.34
2. Songsak	24387	67.65	99.69	0.12	0.21
3. Resu Belpara	48215	15.65	92.22	7.01	0.77
4. Samanda	19619	52.58	99.81	-	0.16
West Garo Hills	**369877**	**38.33**	**92.56**	**4.20**	**2.93**
1. Resu Belpara (w)	18496	26.59	98.86	0.41	0.73
2. Dadenggiri	49351	36.14	97.43	0.12	2.45
3. Selsella	56430	21.13	97.94	1.22	1.67
4. Rongram	65937	44.54	97.86	0.59	0.55
5. Betasing	41811	28.84	84.60	5.87	9.53
6. Zikzak	40935	36.20	96.02	0.62	3.36
7. Dalu	35103	48.98	97.56	0.84	1.60
8. Chokpot	29597	45.85	79.91	18.58	1.51
9. Dambuk Aga	20682	62.13	75.99	16.06	7.95
10. Rongara	12075	63.18	96.74	0.81	2.45
Meghalaya	**255935**	**20.21**	**90.03**	**3.44**	**6.53**

Source : As in Table- 6.1.

People Involved in Jhum in Rural and Urban Areas

The maximum share of jhum land (99.9 per cent) is located in the rural areas in which 10.4 to 42.8 per cent of the total population is engaged. The jhum land is negligible in the urban areas because non-agricultural activities dominate in these areas (Table-6.5).

Table-6.5: Households Involved in Jhum Cultivation in 1981.

State/District	Total Rural Urban	No. of House holds practi-cing jhum cultivation		No. of Persons engaged in jhum cul-tivation	Percentage engaged in jhum cul-tivation HH persons
	Total	**51720**		**256947**	**19.24**
Meghalaya	**Rural**	**51666**	**(99.9%)**	**256654**	**23.45**
	Urban	**54**	**(0.1%)**	**293**	**0.12**
	Total	7029		35540	6.95
Jaintia Hills	Rural	7029	(100.00%)	35540	10.75
	Urban	-		-	-
	Total	3251		14902	9.53
East Khasi	Rural	3251	(100.00%)	14902	10.39
Hills	Urban	-		-	-
	Total	3629		19207	11.89
West Khasi	Rural	3629	(100.00%)	19207	12.18
Hills	Urban	-		-	-
	Total	28877		141786	38.33
East Garo	Rural	28823	(99.8%)	141493	42.82
Hills	Urban	54	(0.21)	293	0.74
	Total	8935		46512	34.06
West Garo	Rural	8935	(100.00%)	46512	35.17
Hills	Urban	-		-	-

Source : As in Table - 6.1

Area Under Crops

Rice, oil seeds and cash crops are the main crops under jhum cultivation in Meghalaya. Maximum areas, 38 to 71 per cent in the different districts is utilized for cultivation of rice, only 8 to 16 per cent for other food grains, 0.3 to 8.0 per cent area under in oil seeds and 14 to 40 per cent for cash crops. Rice, Maize and Millets are the major crops in Jaintia and East Khasi Hills and West Khasi Hills. In addition, wheat and small millets are also grown in East and West Garo Hills. Among cash crops, Potato is grown maximally 25 to 73 per cent in Jaintia, East Khasi and West Khasi Hills, however, only 1.1 to 2.00 per cent in East and West Garo Hills. Tapioca, Jute, Mesta, Cotton and Banana are the major cash crops in East and West Garo Hills. Sesamum Rape and Mustard are the major oil seeds in different districts of Meghalaya. Generally, mixed cropping and crop rotation are practised.

Mixed Cropping : In the Kharif season, the practice of raising a variety of crops as a mixture is common in jhum land. The combination of mixed cropping are :

Paddy-Pumpkin-Maize-Cotton-Yam-Cucumber-Water-melon and other vegetables in *kharif season*. In the *Rabi* season, Beans, Cabbage and Mustard are grown in the same plot of land.

Crop Rotation : Crop rotation is the important aspect of cropping pattern. The system of crop rotation is still traditional. In late March or early April, the jhumias cultivate mostly vegetables and Maize, such as Pumpkin,

Cucumber, Brinjal and Chilies for the cultivation of cash crops like Sesamum, Tobacco and Ginger, the farmers select a fertile plot of land, these plots are not mixed with other vegetables and crops.

Average Crop Production

Maximum 1137.17 kg. per household paddy, 1009.3 kg per household ginger and 1546.3 kg household fruits is produced in jhum cultivation. Production in permanent cultivation is much higher than in jhum cultivation (Table-6.6). Crop production is too lower compared to settled agriculture in other part of country.

Table-6.6 : Average Quantity of Crops Produces under Jhum and Permanent Cultivation

(Figs. in kg. per household)

Crop	Jhum Cultivation	Permanent Cultivation
Paddy	1137.17	2320.85
Maize	171.71	339.55
Other Cereals	236.10	284.32
Mustard	83.33	341.32
Other Oil Seeds	75.77	-
Potatoes	887.83	1588.41
Other Vegetables	182.96	135.00
Other Fruits	1546.25	-
Ginger	1009.70	1314.29
Chilli	170.26	1170.59
Turmeric	228.72	-
Mesta	400.93	-
Jute	75.00	268.67
Cotton	157.59	-
Tapioca	544.42	-
Sweet Potato	518.75	373.75
Arum & Other Roots	474.44	-
Any Other Crops	45.71	1237.50
Other Pulses	-	202.86
Tobacco	-	41.57
Total	2089.94	2685.68

N.B. : Since an average 1.00 ha land is under possession of each household, the production will be 1137.17 Kg./ha.

Source : As in Table-6.1.

Causes of Low Production

The energy inputs in jhum comprises two parts: (1) energy inputs by man (2) natural energy inputs through biological process and by harvesting solar energy. The energy inputs by man may comprise two part, viz., labour input and energy input by use of chemical fertilizer or machines which largely depends on fossil energy. In case of jhum, labour input is largely for felling of trees, burning the debris, dressing the fields, sowing, protection and harvesting in which protection forms a major component.Jhum cultivators either do not use improved seeds and chemical fertilizer or very little compared to that of the cultivators in permanent lands (Table - 6.7). Table-6.8 summarises the use of seeds, green manure and fertilizer in jhum and permanent cultivation. Improved seeds green manure, cow dung and chemical fertilizers are used more in permanent cultivation.

Minimal use of improved seeds and fertilizers are either due to lack of knowledge, non-availability of technological inputs, people being hesitant about their benefit (Table - 6.7).

How to Improve Economies of Jhumias

The most preferred solution to the problem of shifting cultivation is to change the economic base of the community by providing them some agricultural land. Under this scheme plain land can be easily brought under cultivation, perhaps with some investments. In some places, it is possible to locate extensive plain areas. In case of slope land, terraced fields need to be prepared. On sites of new settlements away from their existing habitations, it is necessary to establish new colonies.

A typical colonisation scheme comprises allotment of agricultural land, provision of financial assistance for the development of lands, purchase of bullocks and implements, supply of seeds for one or two seasons, subsistence allowance for a period of six to twelve months, and sometimes

Table-6.7 : Households Not Using Improved Seed and Fertilizer and Its Causes.

Persons	Per cent Household			
	Improved Seeds		Chemical Fertilizer	
	Jhum	Permanent	Jhum	Permanent
% Households not using improved seeds & fertilizer	100.00	95.33	94.83	82.87
(A) Causes				
1. No Knowledge	13.93	13.61	15.40	8.98
2. Not Interested	26.97	25.59	33.41	32.99
3. Not Certain About	13.72	13.43	13.74	13.15
(B) Benefit				
4. Short Supply	0.67	2.36	-	2.30
5. Not Available	36.40	32.67	27.25	32.15
6. Lack of Funds	5.84	11.07	6.16	8.55
7. Others	2.47	1.27	4.03	1.88
Total	100.00	100.00	100.00	100.00

Source : As in Table- 6.1.

Table-6.8 : Percent Households Using Improved Seeds, Green Manure, Cowdung and Chemical Fertilizers in Jhum and Permanent Cultivation.

Extent of Use	Jhum Cultivation				Permanent Cultivation			
	Impr. Seed	Green Manure	Cow Dung	Chemical Ferti.	Impr. Seed	Green Manur	Cow Dung	Chem. Fert
Entire Area	-	2.25	5.62	5.17	0.69	6.57	15.74	3.46
Half or More	-	0.90	-	-	1.04	9.00	3.64	7.27
Less than Half	-	-	-	2.94	0.87	9.69	6.40	
Not Used	100.00	96.85	94.38	94.83	95.33	83.56	70.93	82.87
Total	100	100	100	100	100	100	100	100

Source : As in Table-6.1.

Table-6.9 : Household Types Practising Jhum, Permanent and Both Jhum and Permanent Cultivation

(fig. in percent)

Household Group	Jhum	Permanent	Jhum & Permanent
1. Scheduled Caste	-	85.71	-
2. Khasi & Jaintia	8.88	53.05	31.47
3. Garo	44.24	21.41	15.53
4. Hajong	-	84.62	-
5. Rabha	-	100.00	-
6. Mikir	-	33.33	53.33
7. Kachari	-	100.00	-
8. Biate	80.00	-	20.00
9. Vaipheh	100.00	-	-
10. All Sch. tribes	27.58	37.00	22.93
11. Others	-	61.11	-
Total	25.31	39.17	21.04

N.B.:- The remaining are non-cultivating households.

Source : As in Table-6.1.

establishment of a colony with residential houses, schools, drinking water and other social amenities. Agricultural development programmes inevitably tend to concentrate their efforts on the factors that seem most crucial to crop production such as popularization of high yielding varieties of key grain crops, the proliferation of irrigation systems and the introduction of fertilizer and other inputs. Jhum cultivators in Meghalaya are not using improved seed either due to lack of knowledge, non-availability of improved variety seeds or they are not sure about the benefits. Even in settled agriculture, only 5 per cent cultivators are using chemical fertilizers Because of the size of land holdings and the constraints of yield, shifting cultivation tends to inhibit the productivity of the land and can not absorb other resources which contribute to productivity. Researchs are, therefore, needed to determine the carrying capacity of land to give a crude estimate of the number of people that a given type of land can support with a given technology. Stress should be laid on proper resource management including horticulture and animal husbandry in conjunction with shifting cultivation.

In addition to introduction of high yielding food crops, trees satisfying some of the farmers structural, animal feed, fuelwood and other requirements need to be planted.

Change in Ownership Pattern

Community ownership of resources with some local variations of detail has been generally the basis of jhum economy. The contact with other systems where private ownership is prevalent has induced some changes among the shifting cultivators as well. In many areas, as a result of growing pressure of population, shifting cultivators are adopting to settled cultivation to varying extent. In Meghalaya, the cultivators are moving towards permanent cultivation, however, Vaipheh, Biate are still sticking with jhum cultivation. Schedule Castes, namely, Hajong, Rabha and Kachari have maximally adopted permanent cultivation. However, Khasi,

Jaintia, Garo, Mikir, Biate and all schedule tribes are doing both jhum as well as permanent cultivation (Table- 6.9). In most cases, the individual has only limited ownership right over his fields; he can not transfer the land. Even a developed permanent field reverts to the community if the individual discontinues cultivation or leaves the village.

Programmes for Improving Shifting Cultivation

In most areas of the North-Eastern India, a greater span of time for fallow are required before crops can be successfully cultivated again. Moreover, the fallow period of jhum varies according to soil types and ecological zones. There are areas where fallow periods have been reduced to less than two years under such conditions, farmers face problems of soil degradation, loss of fertility, reduced yields, rampant weed growth and infestation by pests. As a result,there is an urgent need to improve or develop new efficient systems for sustained yields which are scientifically sound, economically viable, and culturally acceptable.

Such a system should have following basic ingredients :

1. Systems should improve on the capacity of shifting cultivation to accommodate population growth.
2. Systems should be suitable for farmers especially marginal and small scale farmers.
3. Systems should be environmentally sound, and at the same time supply farmers with their requirements of food and cash.

Researches

Researches and problem specific studies should be conducted in order to collect information and to understand the functioning of the system. This is also required for mid-term correction and will help for further improvement of the working and acceptability of the system. The following aspects must be emphasizes in detail :

1. The farmers environment, including agroclimatology, soil, vegetation, and pests.
2. Existing farming and productive systems.
3. Commodities produced, including trees, crops and livestock.
4. Resources and its management practices.
5. Input/output ratio for energy and materials.
6. Roads, transportation facilities, marketing and pricing systems.
7. Identification of social and physical Constraints.

Training and Extension

The collaborative programmes need to familiarize farmers with the results of its researchs and to help them develop the new skills they require. Some of them can be accomplished by the involvement of farmers in the experimental projects described above. Research extension usually involves interaction between farmers and extension workers, which provides the feedback to researchers. The process is strengthened when extension workers share the same experiences as the farmers.

Publications

Publications can also give farmers, extension workers, and policy makers greater familiarity with alternatives to and improvements in shifting cultivation. The news media can also be used in campaignings which make people aware of common practices that cause soil erosion and environmental deterioration.

FAO Role in Improving Shifting Cultivation

FAO (Food and Agriculture Organisation of United

Nations) has played a unique role in the past, and is in a position to make particularly important contribution. FAO is currently providing assistance for improving shifting cultivation in following ways :

1. It has contact with many regional and inter-national institutions involved in agricultural research and development.
2. It promotes exchanges of experiences and practices among countries through technical co-operation among developing countries (TCDC).
3. It informs policy makers, through the FAO statutory bodies and other fora to introduce the action needed for improvements and alternatives to shifting cultivation.
4. It develops projects at the national level to demonstrate improvements in shifting cultivation.

7

Land Ownership Pattern in Meghalaya

B. K. Tiwari
Centre for Eco-Development,
North-Eastern Hill University, Shillong
and
S. K. Shahi
Regional Centre, N.A.E.B.,
North-Eastern Hill University, Shillong

Introduction

Meghalaya, the 21st state of Indian union, lies from 25°47' to 26°10' North latitudes and 89°45' to 92°47' East longitudes. It is bounded by the state of Assam to the North and East and Bangladesh to the South and West. It is one of the hill states of North-East India. It comprises 5 districts, namely, West Garo Hills, East Garo Hills, West Khasi Hills, East Khasi Hills and Jaintia Hills with an area of 22429 sq. kms. It is predominantly a tribal state. According to 1981 census, the total population of Meghalaya is 1335819 of which tribal population is 1076345, i.e., 80.52 per cent. The major tribes that inhabit the state are the Khasi-Jaintia in the Khasi and Jaintia Hills districts and Garos in the Garo Hills district.

The climate of Meghalaya is neither too hot in summer nor too cold in winter, but it varies regionally due to uneven hilly terrain. The Northern and Southern slopes as well as the Western border lands of the state are warm in winter due to lower altitudes of these areas. But the central plateau is cold during this season. The average temperature of Meghalaya is 22°C in the summers and 10°C in winters. The average annual rainfall is 400 inches.

The economy of Meghalaya is predominantly agriculture based. Agriculture is a major activity in the state, both in terms of employment, resource and its contribution to total production. The state has a major portion of its workforce engaged in agricultural sector. According to 1981 census, 72.54 per cent of its total main workers and 83.79 per cent of total rural main workers are directly engaged in this sector. This figure shows a high population pressure on land like other parts of the country; the rapid growth rate of population and population density within a few decades also indicate an increasing pressure on land in the state. Since the population pressure on land is increasing, the study of land management, landuse and land ownership patterns have become more important.

The present paper attempts to study land ownership pattern in Meghalaya in relation to its physical and socio-cultural background.

Land Ownership

The basic concept of land ownership, according to Davis (1978), is that of tenure, i.e., the right of capacity to have and hold land for certain uses. According to Hoebel and Frost (1976) land tenure deals with the interrelated network of rights, duties privileges and immunities which individuals and groups within a society observe in relation to land resources. The U.N.O. report on progress on land reform (1970) has given the definition of ownership of land as :

> " ... the right to use land, together with the right to transfer that right to others. Also both of these rights are more or less circumscribed by national or local laws, so that exact context of 'ownership' varies from society to society".

The pattern of land ownership plays an important role in landuse and agricultural development planning. Therefore, the better understanding of the form of land ownership and related laws give an insight into the problems faced by our present day society specially in the field of agricultural development and landuse planning. The form of ownership and accompanying laws are quite different in Meghalaya than those of other parts of the country. Even in Meghalaya, they differ from place to place and from one tribe to the other. Land ownership system in Meghalaya is broadly of two types, *viz.*, *Riotwary* and Customary land systems which vary from district to district. The principle of *Riotwary* system is that the Government deals directly with the actual land holders without the intervention of intermediaries. Customary system is totally different from *Riotwary* system. According to U.N.O. report on progress on land reform (1970), in Customary land tenure,

> "the right to use or to dispose of use rights over land rests neither on the exercise of brute forces, nor on statutory evidence of rights guaranteed by Government statute, but on the fact that they are recognized as legitimate by the community, the rules governing the transmission of these rights being usually explicit and generally known, though not normally recorded in writing".

On account of variations of the physical features in the state, there are three important tribes, *viz.*, Garos, Khasis and Jaintias inhabiting in the three different hills and have different socio-economic traditions. Therefore, the land ownership patterns also vary accordingly. The detail study of the land ownership may be described in the following manner.

(a) Ownership Pattern Among Garos

In Garo Hills, *Riotwary* system is prevalent in the *mauzas* which have plain lands and Customary system in the hill *mauzas*. The existing pattern of land ownership in Garo Hills is related to the types of cultivation practised in the areas. Among the Garos, shifting cultivation practising private ownership, is restricted and everybody has a particular unsufructuary right on land, i.e., they have the right to use land without possessing an actual title of ownership. But by the arrival of wet-rice cultivation which makes private ownership of the land essential, a change in the pattern of land ownership has emerged and private ownership has increased. In many cases, private ownership is associated with permanent holdings. But unlike other peasant societies in the plains, the Garos, even if they possess such a sort of permanent holdings on land, they cannot sell or transfer it to the outsiders (non-natives) if the holding is inside one's *Aabhang* land. It means the land should be always with the native community and is a common property of the people.

The village territory under a chief is known as *Aabhangland*. The *Aabhang - Nokma* who is the heiress daughter of a family and is selected from the oldest family among the *Machongs* initially belonge to two super clans, called *Chatchi, viz.*, Sangma and Marak. The *Chatchi* develops strictly as the descendants of the same mother who lives in geographically delimited areas, are called *Machongs*. The *Abhang-Nokma* is the actual owner of the *Aabhong* land but her husband (*Nokma*) acquires the right of managing the whole property of the village. The *Nokma*, however, has no power of selling any part of the territory to another village or man without the permission of the respective *Maharis* in the village. The institution of *Mahari* is consists of the members closely related through common motherhood and collectively responsible for the conduct and security of the members and protection of the family property (Bhattacharya 1985). In the event of a person, who desires to cultivate land within the boundary of another *Nokma,* he would give an undertaking to pay a trifling amount of the produce to the *Nokma* of the latter at the time of harvesting. Such kind of payment is known as *Harshel. Harshel* may, therefore, be considered as a kind of rent, which varies in different localities. A *Nokma,* however, does not take *Harshel* from his own village members. The *Aakhing* or *Aabhang Nokma* in matter of cultivation is like a common villager. But as regard to the forest resources in the *Aabhang* land and in distribution of plots, he has certain prerogatives (Banarjee 1972). Titiev (1944) expresses,

> "the village chief is the theoretical owner of all his village's land; these lands are divided among the clans residing in his village".

Among the Garos, a user right on land does not necessarily confer any right to property. The wild and semi-wild trees are regarded as community property and everybody has the right to use it, if it is not grown inside one's plot. If a man plants tree, it belongs to him and he may retain the right of ownership of the tree which he has planted even after forfeiting his user right to the land. If a man decides to allow some of its plots to lie fallow, no other person may use them without the specific permission of the earlier user as well as *Nokma* of the village.

(b) Ownership Pattern Among Jaintias

In Jaintia Hills, there is no rigid, uniform or regular systems worth the name with the exception of those under the State Forest Department and Local District Council. However, the land can be placed under following categories (Lahiri 1979) :

(i) ***Ri Kynti*** or private land : Total area under this category is 7266 acres and they are known as *"Biniaj"*.

(ii) **Raj Land** : All Raj lands which have been brought under wet paddy cultivation are subjected to payment of annual revenue. The revenue is fixed according to the fertility of land which is classified as *Hali*; (land class I, II and III). The revenue varies according to these three classes from 50 paise to 87 paise per year. There are altogether 28683 plots covering a total area of 19709 acres in Jaintia Hills district. But in Khasi Hills, no such revenue exists.

(iii) **Land under *Zamindars*** : In the border areas (War area) of both Khasi Hills and Jaintia Hills districts, there are some big land holders who are locally known as *Zamindars*. There is no record of the number of such *Zamindars* nor the areas of land held by them. Generally the *Zamindars* lease out their land on the basis of rent in cash or in kind.

In the Jaintia Hills, *Doloi* elected from among the male members of the original clan or clans (*Kurs*) who were the earliest settlers of Raj or *Illaka*, in the pre-British period had the power and authority to grant license and *pattah* for cultivation of a newly terraced wet rice cultivation. This privilege remains with the *Doloi* even today. The individuals granted such new *pattah* for their wet rice cultivation have to pay a yearly fee to the *Doloi*. Such payments are shared by the *Doloi* and his deputies. In some villages where abundant forests for the jhum cultivation in the Raj lands are available, the village Headman convinces the should take place. After the selection of the place, the village Headman with the help of the other village officials distributes the land to all the families and see that all get their due share. The jhum land is divided among the cultivators according to their farming abilities (Gassah 1980).

(c) Ownership Pattern Among Khasis

In Khasi custom, there are many laws related to land and its accompanying ownership. Therefore, the classification of land ownership is extremely complicated. On the basis of three basic criteria of owner of land, 4 types of land ownerships have been identified in Khasi Hills (Haloi 1984).

These are :

(i) private ownership,

(ii) group ownership,

(iii) community ownership, and

(iv) the Government ownership.

(i) Private Land : The land over which the single owner or individual family owners enjoy all the rights at will and in no case is subjected to the control of any superior authority, can be defined as the private land. Bareh (1976) describes that the owner of these land enjoys all the rights over land. The owner can sell or buy at will. No owner needs to pay any revenue to any authority. The Land Reforms Commission (1974) clarifies that the State Assembly or any other Assembly has no control over the private land except with regard to the settlement of dispute about them if and when they are brought to the Assembly of the village. Roy (1934) states that the land property may be of ancestral or purchased type. The lands coming under this category are:

(i) *Ri-Nongtymmen,* (ii) *Ri-Maw,* (iii) *Ri-Khurid,* (iv) *Ri - Bitor,* (v) *Ri - Dabhol,* (vi) *Ri-Shyieng,* (vii) *Ri-Phriang,* (viii) *Ri-Spah,* (ix) *Ri-Langdung,* (x) *Ri-Samla* and (xi) *Ri-Nongmei Nongpa* .

Thus, the *Ri Kynti* or private land system endows with absolute possession of land. According to the land reform commission, *Ka-Khadduh* (youngest daughter of the family) gets more share in comparison to her other sisters. If the parents of the family do not have any daughter, then the family brings a *Ka-Khadduh* from the sister's family immediately to the mother. In absence of them, the family prefers a cousin daughter from the mother's side. Through *Pynbem* (a land distribution ceremony) land is distributed among the sisters of the family. The mother with an uncle or a brother living in her house, or with husband or alone may apportion the land among her children. In case of her (mother's) death before apportion, the *Ka-Khadduh* by herself can not dispose the property. And when mother is no longer alive, *Ka-Khadduh* as the custodian of the ancestral property is to be assisted by her elder sisters or brothers in its management. A purchased plot of land also becomes an ancestral property of the family just after one or two generations.

(ii) Group Land : The land over which the owner's rights, especially the transferable right, are subjected to the control of group-assembly, can be defined as the group land. The group may consist of a branch of a clan or a group of clans having assembly individually. The members of each and every group are the owners of these land. Each and every group has own *Durbar* (Assembly). As Chowdhari (1978) describes, all members of the concerning group enjoy use-right, occupy right and heritable right at will over their land but no one is entitled to sell the land individually. According to the Land Reform Commission, the right of transfer of land rests upon the decision of the concerned assembly of the group. Gurdon (1975) says that no one need to pay any land revenue to any authority and, thus, it is also applied to the tenants of *Ri-Kynti*.

Ri-Kynti, belonging to a clan, which has not been divided, is known as *Ri-Kur* (Roy 1934). The control and management of these lands rest upon the *Durbar-Kur* (clan assembly) where the eldest uncle of the clan acts as the head. In absence of suitability of the uncle, the eldest brother or another brother who is considered suitable, is elected as the head of the *Durbar-Kur*. The *Durbar - Kur* takes decisions also on the partition (if it is desired to be done) of *Ri-Kur* of a clan among its various branches.

After the division of *Ri-Kur* among the branches of the clan, the branches form their own *Durbar*. According to the report of Land Reform Commission, the *Durbar* with the eldest uncle of the branch as head acts as the authority of control and management. The branch assembly and the *Durbar-Kur* work in the same way. The work of the clans assembly is also the same as the other two assemblies. The following various categories of land come under the group land. These are :

(i) *Ri-Kur,* (ii) *Ri-Seng,* (iii) *Ri-Khain,* (iv) *Ri-Durbar,* (v) *Ri-Lyngdoh,* (vi) *Ri-Iapduh* (if owned by clan), (vii) *RiSyiem,* (viii) *Ri-Sniah,* (ix) *Ri-Kut,* (x) *Ri-Law Sumar*(if owned by clan) and (xi) *Ri-Bam Syiem.*

(iii) Community Land : The land which can not be transferred by users or even by the community assembly, but the other rights are subjected to the control of the community assembly, come under community land. A community may consist of one clan or more than one clans. It may cover single village or many villages. Each and every member of the community can enjoy use right, occupancy right, and some times even heritable right, but after having the approval of the community assembly only.

According to the report of Land Reform Commission for Khasi Hills 1974, every member of the *raid* means community has the use-right and occupancy right over *Ri-Raid* (community land). No one has to pay any land revenue for enjoying these rights. If one has to pay any amount, then it is generally for the improvements of land which the community would have made on the land or for services rendered in its use and cultivation. Roy (1934) states that no one claim more land than what one actually can occupy for cultivation purposes or for any productive use. Any one of the community can use and occupy the vacant plot of community land after having the approval of *Raid-Durbar* (community assembly). Chowdhury (1978) explains that the *Raid-Durbar* is the authority of the allocation and distribution of community land. However, authority can sell community land.

Community land is not heritable and transferable. Of course, exceptions are there. Roy (1934) explains the situation like, if one fails to use and occupy the land after three consecutive fallow years, automatically it goes to the community assembly and the assembly allots use-right and occupancy right to others. But if the occupant maintains its use, the person can inherit it to his descendants, but in no case he can enjoy the rights of the transfer. According to Land Reform Commission, if one has made permanent improvement (like in the form of cultivation of permanent crops and plants like fruit trees or converting the land into wet paddy fields or by making permanent building etc.), he may be endowed with transferable rights. Of course, a person loses all rights if he leaves the plot or let it become jungle growth. The community assembly decides the period after which the property may be treated as having revert back to the community. Only a Khasi of the same community can have the use right and occupancy right over community land. A Khasi from another community also enjoy these rights, if he submits himself to the jurisdiction of the community and has been accepted and recognised as one belonging to it by the community assembly. The report of the Land Reform Commission for Khasi Hills (1974) states that except *Durbar-Hima* (State Assembly) no Khasi administrative body has the right over community land to allow the use-right and occupancy right to a *Mynder-Ri* (Non-Khasi) by issuing *pattas* and levying and realising land revenue or land rent from him. The following categories of land come under the community land :

(i) *Ri-Shnat,* (ii) *Ri-Kuna,* (iii) *Mynsain,* (iv) *Lamsain or Somraid or Ri-Pud,* (v) *Ri-Lynter,* (vi) *Ri-Law Kyntang,* (vii) *Ri-Law Lyngdoh,* (viii) *Ri-Law Niam,* (ix) *Ri-Law Sumar,* (x) *Ri-Law Adong,* (xi) *Ri-Law Shnong,* (xii) *Ri-Shnong* (sometimes), (xiii) *Ri-Ialah Mukotduma,* (xiv) *Ri-Umsnam,* (xv) *Ri- Bamlang,* (xvi) *Ri-Aiti Mon Sngewbha,* (xvii) *Ri-Phlang-Ribam Duh* and (xviii) *Ri-Diengsaiduebghub.*

Among these categories some are special land kept by the community for specific uses. The forest land kept for religious and medicinal purposes are: (i) *Ri-Law Kyntang,* (ii) *Ri-Law Niam,* (iii) *Ri-Law Lyngdoh* and for the purpose of firewood, timber, water etc. are : (i)*Ri-Law Adong*, (ii) *Ri-Law Sang*, (iii) *Ri-Law Shnong* and (iv) *Ri-Law Sumar*.

(iv) Government Land : The land which has been leased, purchased or acquired under various rules issued from time to time by the Government can be defined as Government land. The Land Reform Commission for Khasi Hills states that the Government of India has taken the land in the Khasi Hills under various rules issued by the Government. According to the related rules, the Government enjoys all the rights over land. The following categories of land are under Government land: (i) Exchanged land, (ii) Purchased land, (iii) Leased land and (iv) Acquired land.

As the Khasi, Jaintia and Hill *Mauzas* of Garo Hills districts have not yet been cadastrally surveyed, no record of land ownership has been prepared by the Government. So, it is very difficult to say the actual area under different types of land owned by various authorities in this region.

In general, the wet-rice fields which are permanently cultivated are basically under the private ownership in Meghalaya. But the jhum land can be either under private ownership, group ownership or community ownership. The forests are under group, community or public ownership. Forests which are under the public ownership the villagers also have some use rights over these forests (Haloi 1984).

Within the past few years, the new agrarian relations have been developed in Khasi-Jaintia Hills of Meghalaya. The increasing trend in private ownership has been found and a good portion of land is now under private ownership (Sen 1985). In some places, the village headman and the heads of different land owning assemblies are also overlooking the customary laws and helping to increase private land ownership due to their individual interest. A 50 years old report states that, in the areas of Khasi-Jaintia Hills, much of the community land (*Ri-Raid*) was converted into private possession (*Ri- Kynti*). In the areas situated in the surroundings of Shillong town where land fetches high prices, certain village headmen claimed that the *Raid-Durbar* has a right to sell *Raid-lands* and give these land to people as *Ri-Kynti* (David Roy 1934). It is to be remembered that the *Ri-Kynti* owners known as *Bahhraws* are considered as aristocrats. These aristocratic clans always owned better and more fertile land of the village and mostly waste and less fertile lands were distributed among other clans known as *Khian-Kur* or Commoner's clan (Pakem 1984). A number of case studies conducted in recent years have also shown a concentration of land holdings under the ownership of a few families in a village with consequent rise in landlessness and share cropping (Vincent 1976, Lahiri 1979, Blah 1981). As Lahiri states that there are share-croppers or *Adhiars* in a large number in Meghalaya. According to the report on Socio-Economic Survey conducted in Khasi and Jaintia Hills (1966-67), there is about 4 per cent increase in the number of share croppers over a period of 15 years (1951-66).

It is also indicated by the Survey Report that the land is concentrated in a few hands most of whom do not cultivate themselves. A typical case is the village Hat Mawdon in Mawsynram Community Development Block wherever eleven out of fifteen families are found to be landless and were share croppers in the land owned by the other 4 families (Dutta 1981).

According to the Khasi custom and the logic of community ownership of land suggest that a family can hold land only in the village where they reside. But there has been transfer of land between persons who are residents of different areas also. In the Mynsain village, out of 30 families only 6 families have land of their own for cultivation of paddy and the major portion of paddy growing land is owned by a person who is not a resident of that village (Lahiri 1979). The situation negates the hither to believed motion that land in the Khasi and Jaintia Hills are under community ownership (Kar 1981). The situation in Garo Hills is not different either.

Now-a-days, traditional ownership is being replaced by private ownership. In Garo Hills, it is observed that through various development schemes the State Government also sponsored such fashions of community ownership. For example, all terrace plots initially prepared and financed by the Government were subsequently allotted to different households (Kar 1981). Thus, the egalitarian system of community ownership is being replaced by private ownership with the development of societies in the region.

There are many advantages and disadvantages of community ownership. The community ownership among the cultivating communities generally places difficulties in the way of adopting improved farming and landuse, in which every one tends to be regarded as equal, which discourages the individual initiative. The land can not be used as security for development loans, land use planning, farm planning and the introduction of better farming system are rendered difficult. But community ownership of land and communal responsibility in agriculture offer its members security. The community ownership system provides an excellent foundation for the co-operative farming system towards which many parts of the world are moving. In the present degrading environmental conditions in the hills of Meghalaya, the community ownership system is also preferred because the hill tribes are accustomed to community way of living.

Thus, it can be concluded here that the pattern of land ownership and the classification of land is positively related but it a task to understand the subject because of non-availability of correct and required data. However, the present study highlights the facts that, for better management of land, it is necessary to have a clear and simple system of land classification and ownership. The complex pattern of ownership prevailing in this state are often found difficult for legal interpretations. One of the major impediments of the existing land laws and ownership practices are not conducive for the financial institutions like banks and co-operative societies and, consequently, the rate of development is very low. In this connection, it can be said obviously that a systematic study of the land ownership pattern in relation to customary laws must be undertaken. Therefore, for a faster agricultural development in the region, a systematic study of the land laws and the land ownership patterns is necessary to simplify and standardize the land ownership patterns and its related customary laws.

References

1. Bannery, N. K. (1972): An Appraisal of Shifting Cultivation in India, in L.P. Patel (e d.) : *Agro-Economic Problems of Tribal India*, Delhi.

2. Bareh, H. (1978) : *The History and Culture of the Khasi People*, Acharya Praffulla Chandra Road, Calcutta-4.

3. Bhattacharya, J. B. (1985) : Social Formation in Precolonial Garo Society Problems and Process, in S. K. Chattopadhyaya (ed.)*Tribal Institution of Meghalaya* , Spectrum Publications, Guwahati.

4. Blah, R. P. (1981): Past and Present Agrarian Social Structure, Paper Presented in a Seminar on *Land Relations in the Hills of North East India, NEICSSR*, Shillong (unpublished).

5. Chowdhuri, J. N. (1978) : *The Khasi Canvai*, Shillong - 1.

6. Devis, K. P. (1976): *Land Use*, Mac-Graw Hill Book Co., Delhi.

7. Dutta, B. B. (1981): Land Relation in Khasi Hills- A Study in Ownership and Alienation, Paper Presented in a Seminar on *Land and Land Relation*.

8. F.A.O. and I.L.O. (1970): *Progress in Land Reform* (Fifth Print). U.N.O., New York.

9. Gassah, L. S. (1980) : *Traditional Leadership in Jaintia Hills: The System of Political Administration in the Past and Present*, Unpublished M.Phil Thesis, Department of Sociology & Anthropology, NEHU,Shillong.

10. Gurdon, P. R. I. (1975): *The Khasi*, Cosme Publica-tion, Delhi-6.

11. Hobel, E. A. and Frost, E. C. (1976): *Cultural and Social Anthropology*, Mec-Graw Hill Co., New York.

12. Haloi, K. (1984): *Concepts of Landuse and Owner-ship in East Khasi Hills*, Unpublished M. Phil Thesis, Department of Geography, NEHU, Shillong.

13. Kar, P. C. (1979): Land Relations and Changing Agricultural Practices, in B. Dutta Ray (ed.): *Agriculture in the Hills: A Case Study of Meghalaya* , NEICSSR, Shillong.

14. Lahiri, B. (1979): Land Relations and Changing Agricultural Practices, in B. Dutta Ray (ed.): *Agriculture in the Hills: A Case Study of Meghalaya*, NEICSSR, Shillong.

15. Momin, P.G. (1977): Land Use Pattern in Meghalaya, Paper Presented in a Seminar on *Resource Base & Problems of Regional Development of North-Eastern Region of India*, Deptartment of Geography, NEHU, Shillong (Unpub.).

16. Pakem, B. (1984): State Formation in Pre-Colonial Jaintia, *The NEHU Journal of Social Sciences*, Vol. II, No. 3.

17. Roy, D. (1934) : Land Tenure in the States, in *Sir Keith Cantlie's Notes on Khasi Laws*, Ri Khasi Press, Shillong.

18. Rymbai, R.T. (1974)(Chairman) : *Report of the Land Reforms Commission for Khasi Hills.*

19. Sen, S. (1985) : *Social and State Formation in Khasi and Jaintia Hills*, B.R. PublishingCo., New Delhi.

20. Vincent, K. (1976) : A Peasant Economy in Tribal Area - A Case Study of Bhoilymbong Village in the Khasi Hills of Meghalaya, in T. Mathew (ed.): *Proceedings of the Seminar on Tribal Economy*, Deptartment of Economics, NEHU, Shillong.

[illegible]

[illegible] (19[illegible]) [illegible]

[illegible] (19[illegible]) [illegible]

[illegible] (19[illegible]) [illegible] Land Relations and Changes in Agricultural [illegible]

[illegible] (19[illegible]) Land Relations and Changes in Agricultural [illegible]

[illegible] Meagher, RG (1977) [illegible]

[illegible] (19[illegible]) [illegible]

[illegible] (1956) Land Tenure in the [illegible]

[illegible]

[illegible]

8

Landuse Patterns in Meghalaya

Surendra Singh
Department of Geography,
North-Eastern Hill University, Shillong

Introduction

Land is a piece of resources and the study of landuse has great significance for the proper use of land resources. The availability of land resources, natural as well as human and their potentials form the complex nature of interactions between man and nature to satisfy human needs. Land-man relationship is always influenced by various geographical factors which evolve the landuse pattern of an area or a region. Therefore, landuse characteristics are directly related to the physical attributes of land, namely, slope, relief, climatic conditions, geological structure etc. and socio-economic and cultural factors like land tenure system, size of landholdings, applicability of input factors, effects of production prices and traditional systems of farm operations. Landuse practices are also the reflection of the application of modern technology and cultural constraints of the region. For example, the landuse characteristics of the mountain areas of the world are mainly controlled by the physical factors of land, while the agriculturally dominated areas have the strong impact of techno-economic factors and thereby landuse practices are intensified by these factors. At micro-areal level, the processes of landuse are also operated by the accessibility of market centres and transport costs (Thunen, 1933; Viser 1981, 1982).

Indeed, the term 'Landuse' refers to the use of a 'tract' which is generally defined as specific area full of resources, then question arises how man is utilising these resources. It is a fact that the landuse of Meghalaya state is the result of various physical characteristics of land and socio-economic factors. These attributes and determinents of landuse would be studied in detail.

Landuse in relation to Physiographic Conditions

Meghalaya state includes an area of about 22.49 thousand sq.kms. Physiographically, it has its own identity because it is predominantly a plateau land of about 1500 metres height from the mean sea level. The altitudinal variation of the plateau ranges from 500 m to 2000 m approximately. The entire area of the state is rectangular in shape which falls between 25^{0}10' to 26^{0}N. latitude and 88^{0}55' to 93^{0}5' E. longitudes. The Meghalaya state is surrounded by the main plains, namely, the plains of the Brahmaputra valley in the North and in the East and the Bangladesh plains in the South and South-east. Thus, the Assam state is situated in the North and Eastern borders of the Meghalaya state and its Southern and South-western boundaries follow the international boundary of Bangladesh. The contour map of the state which has been prepared by Chatterjee (1968) from one million Topo Sheet, shows that entire state is hilly of which South-eastern parts have steep slope-gradients at higher altitudes of about 2000 m and have less sloppy and undulating land in the central part of the plateau called 'central table land' of Meghalaya. The Jaintia Hills of the East and the Eastern part of Khasi Hills where Shillong is situated are more undulating which follow very high gradients of surface slopes (about 70^0) with the varieties of canyons, gorges and caves where natural vegetation is dominant and most of the share of land is under forests and unproductive categories of landuses. The central table-land of the Meghalaya plateau which covers most of the parts of West Khasi Hills district has smooth topographic features of open river valleys of inter-hill-range plains of gentle slopes where most of the land is under cultivation. In fact, the Westward slopes of the central table-land (i.e. Garo Hills) where eroded hill-tops and wide river valleys can be seen at an average altitude of 1500 m, are used for grazing activity and a significant part of the land is under cultivable waste (Fig. - 8.1).

Intensity and extent of forest land are directly related to topographic features and climatic conditions. In fact, pine forests are mainly found above 1000 m altitudes where the annual temperature ranges from 0°C in winter to 27°C in summer months with the normal rainfall of about 450 mm mostly occuring in summers (Table - 8.1). The central table-land which includes most of the area of East and West Khasi Hills districts is covered by the pine forests. On account of high rainfall with a moderately high temperature in the Southern and Western parts of the Meghalaya, vegetation of these areas comprising most of parts of Garo Hills district and the Southern parts of East and West Khasi Hills districts, is temperate with the mixed forest of hard woods and tall grasses. The area under reserved and protected forests is only 8.5 per cent of the total forest area, however, map of vegetation and forest types also establishes the relationship of forest areas with the altitude of the places. In fact, the reserved forests are situated mostly on the slopes of the main river valleys, while the areas of pine forests are found particularly on the main central table-land which is above 1000 m of altitudes.

The causes of the prevalence of high percentages of cultivable waste and unproductive barren lands are mainly tremendous soil erosion which occurs on the hill-tops and hill-slopes because of deforestation and land clearing for shifting cultivation (Rai and Panda 1983); these being socio-economic causes of land degradation. But the rate of soil erosion and land degradation is direct result of geomorphic features which are the main attributes of the physical characteristics of land. Another attribute is related to climatic conditions, that is the quantity of water run off. With the help of geomorphic characteristics of the mountain land of Ethiopia, Chandra Bhan (1988) calculated the potential soil erosion risk and asserted that the main faces of hill slopes determine the rate of soil erosion and landuse types. Crusts of hill tops (waxing slopes) which are prominent in the upper parts of

Table-8.1: Monthly Mean Temperature and Normal Rainfall, Shillong.

Months	Temperature (°C)		Rainfall
	Maxi	Mini	(in mm)
January	15.5	3.6	25.1
February	17.1	6.4	32.2
March	21.5	10.5	74.0
April	23.8	14.1	179.2
May	23.7	15.5	385.0
June	23.7	17.4	463.1
July	24.1	18.1	416.0
August	24.1	17.8	343.7
September	23.6	16.6	268.2
October	21.8	12.9	135.1
November	18.9	7.7	28.6
December	16.4	4.5	9.9

N.B. : Normal Rainfall figures are the 30 years average for whole of Meghalaya state.

Source : Temperature figures are taken from Rai and Panda (1983).

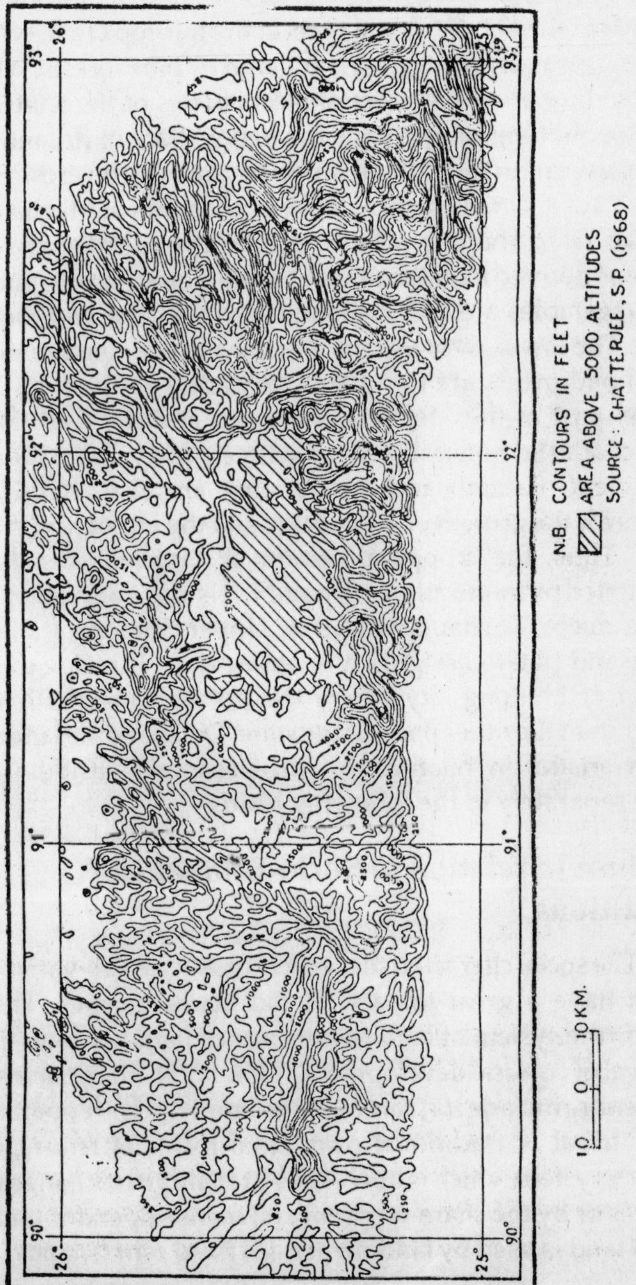

Fig. - 8.1 : Relief Features of Meghalaya Plateau (Contour Map).

Shillong ranges follow the varieties of landuse at different altitudes. Most of the upper parts of the hill-top crust (where convex slopes are markable) is covered by pine forests. While the free-face eroded surface of steep slopes of the Southern parts of Shillong ranges have the major share of its landuse under unproductive barren lands because of the erosion of surface soils. On the debris slopes of the parts of Meghalaya plateau near Shillong which receive the weathered rock material derived from free-face of steep slopes, maize, potato and vegetables are predominantly grown (Rai and Panda 1983). The lower river valleys which have the concave slope called pediments are important for paddy cultivation. The central part of the Meghalaya plateau and the Garo Hills district are the noticeable areas where paddy cultivation is prominent. Rainfall and temperature are also significant factors for the growth of agriculture and the increase of crop-yield. Thus, the cropping pattern of Meghalaya state is dominated by monoculture of paddy. It is also noticeable that in the deep 'V' shape valleys of Upper Shillong Ranges, maize and potato are grown in abundance. It is due to the impact of Shillong city where the potato consumption is higher than the other parts of the state. Details of the landuse characteristics in relation to socio-economic attribute are given separately in the following heads.

Landuse in relation to Socio-Economic Conditions

The social characteristics embrace several sub-variables which have a great impact on the landuse types. These include the system of land tenure, size of land holdings and the level of cultural development (Vink 1975). The systems of land tenure include: (a) common systems which are operated under tribal or traditional communal forms of tenure, (b) tenancy system which is operated by tenants in exchange for services or by the share-cropping, (c) owner-operator under which land is used by land owner, (d) fixed rent tenancy, (e) co-operative holding system under which land holdings are operated by profit organisations, and (f) collective state holdings which are operated by the group of working people collectively (Kostrowicky 1972). Further, landuse is also influenced by various economic factors. Demand of the goods and prices of production are main factors. For example, in spite of sharp topographic features and steep slopes of the river valleys around Shillong city, the crops like vegetables, potato and even maize are grown on commercial basis because of increasing demand of these crop-products in the Shillong urban area. These factors can be interpreted side-by-side in the study of general landuse of Meghalaya state.

Landuse Characterstics

According to the standard classification of landuse which is based on the guidelines of the Ministry of Agriculture and Co-operation, Government of India, New Delhi, the total reporting land is broadly categorised into four categories : (i) Forests, (ii) Land not available for cultivation, (iii) Uncultivated land excluding fallow land and (iv) the Net Area Sown. These categories have been divided into 9 sub- categories which give the detail picture of landuse of the area. Land under various categories and its changing patterns for the Meghalaya state are given in Table- 8.2 & Fig.- 8.2. The impact of socio economic factors on the landuse characteristics of the Meghalaya state is visualised by studying the main landuse categories.

(A) Land Under Forests : It is the most significant category of landuse which accounts for more than one- third share of the total reporting area of Meghalaya state (Table- 8.2). Land under forests was recorded 740 thousand hectare in 1972-73. It was 32.90 per cent of the total area in 1972-73. It is noticeable that it has increased by 3.10 per cent from 740 to 810 thousand hactare during the last 15 years (1972-73 to 1987-88). The reasons behind an increase of forest area in the Meghalaya are as follows :

Table- 8.2 : Changing Patterns of Landuse in Meghalaya

Landuse Categories	1972-73		1987-88		Change
	Area	%	Area	%	%
1. Forests	740	32.90	810	36.0	3.10
2. Area put to non-agricultural uses	50	2.22	90	4.0	1.78
3. Barren & Uncultivable Land	216	9.60	230	10.0	0.40
4. Permanent Pastures & Grazing Land	54	2.40	20	0.90	-1.50
5. Land Under Misc., Tree, Crops & Groves	76	3.38	150	6.7	3.32
6. Cultivable Waste	650	28.90	450	20.0	-8.90
7. Old Fallow	252	11.20	260	11.6	0.40
8. Current Fallow	49	2.18	50	2.2	0.02
9. Net Area Sown	164	7.29	190	8.4	1.11
Total Reporting Area	2249	100.00		100.00	-

Source : Directorate of Eco. & Stat., M. O. A. & I., GOI, New Delhi.

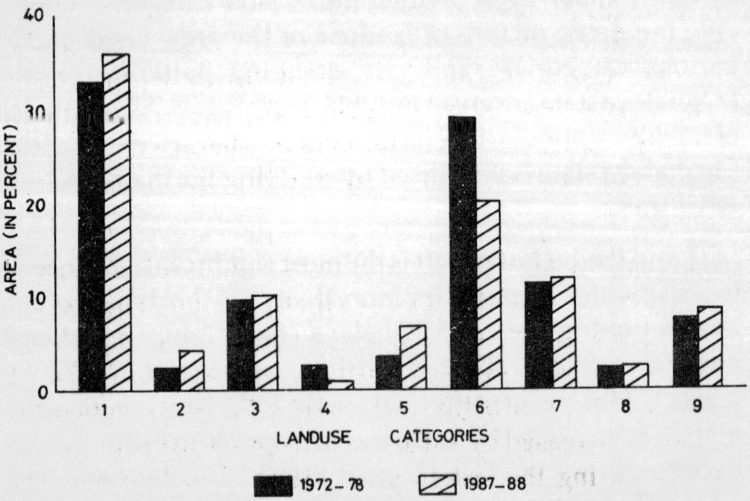

Fig.-8.2 : Changing Landuse Patterns in Meghalaya.

(1) Inspite of more than 90 per cent forest area being under individual tenancy system as recorded unclassified category according to legal status of forests (Table-8.3), the forest area has been increased by 3.1 per cent during 1972-73 to 1987-88. It can be because of general awareness regarding the importance and utility of the forests which is persuaded by the government under the national forest policies. People of the state are aware of the programmes related to social forestry and hence, land under tree crops and groves has also been increased by 3.3 per cent over he last 15 years.

Table- 8.3 : Area Under Forest According to their Legal Status (1984-85).

	Area (000 ha)	%
Reserved Forests	70.6	8.29
Protected Forests	1.6	0.19
Unclassified Forests	779.2	91.52
Total Forest	851.4	100.00

Source : Basic Statistics of N.E. Region 1987, Published by NEC, Shillong.

(2) Indeed, the tribal people are very much fond of the Orchards and Greenaries. They generally maintain kitchen-gardens and tiny plants. As a result, the land under vegetables and groves has increased in the last few years. One of the most important reasons behind increasing land under tree-crops is the increasing demand of the fruits and vegetables. It is a fact that, in the Meghalaya state, the flat land available for cultivation is very less because of hilly topography. Therefore, tree-crops are predominent on the sloppy terrain on which the farmers are practising terrace cultivation.

(3) The decrease in area under shifting cultivation also has a direct impact on the increase of forest land in the Meghalaya

plateau. It is interesting to note that the relief features have a positive relationship with the distribution of the types of forest areas. For example, the areas above 1000 m altitude from mean sea level have pine forests. Therefore, entire parts of the central plateau which includes most of the parts of East and West khasi Hills districts have pine dominated forests. In the valleys of the main rivers, generally in the upper valley areas of Someswari, Dudhnain Umkhri, Umiam and Kopili rivers, the reserved forest areas are markable, although they have only 8.29 per cent share (70.6 thousand hactare) to the total forest area of the state. On the steep slopes of the main rivers, mainly mixed temperate forests are found.

(B) Area Put to No-Agricultural Uses : On account of increasing land under settlements, the area under this category has increased from 50 to 90 thousand hactare in the state during the last 15 years of period. The processes of settlement expansion specially in the urban areas increase the land under non-agricultural uses. In fact, the townships like Tura, Nongstoin, Cherrapunjee, Jowai and even larger villages are in the process of fast population growth and areal expansion. Thus, the area under this category of landuse has increased by 2 per cent.

(C) Land Under Cultivable Waste : General landuse table shows that area under cultivable waste and barren land has decreased by 8.0 per cent from 650 to 450 thousand hectare during the last 15 years. It is noticeable that a significant share of land i.e., 29.0 per cent, is under this category. However, it has reduced significantly. It is due to conversion of waste land into forest area ; in some parts of the state, it has been converted into terrace-cultivation.

(D) Cultivated Land : It includes Net Area Sown and fallow land (other than current and current fallow). Marginal increase of 1.11 per cent may be observed in Net Area Sown. It was only 164 thousand hectare in 1972-73 which has increased upto 190 thousand hectares during the last 15 years. The percentage share of cultivated land is very less on account of physiographic constraints. Going through Table-8.4 which shows the changing cropping patterns, it can be said that paddy is the dominant crop of the state which occupies more than 54 per cent share of the total Gross Cropped Area. The intensive paddy cultivation is practised twice even thrice in a year in the river valleys and on the hill slope terraces. Coarse cereals have the second place in the cropping pattern. There is a marginal decrease in the paddy crop area. In general, it can be said that the cropping patterns of Meghalaya state have not undergone significant changes in its character.

Size of operational land holdings and the system of land tenancy are the main reasons behind the crop dominance and constant cropping patterns. According to Agriculture Census 1970-71 and also 1980-81, the area under various size land holdings was 252.8 thousand hectares (11.24 per cent) in 1970-71 which has increased upto 296.0 thousand hectare in the 1970's. It might be because of increasing area under cultivated land. It is interestng to note that more than 57 per cent of the total land under various size of land holdings was under 1.0 to 3.0 hectare of operational land holding size in 1970-71 which has been reduced to 45 percent during 1970-71 to 1980-81 (Table-8.5, Fig.-8.3). It indicates that concentration of the land holding was towards medium size of 1-3 hectare in 1970-71 which has been diversified towards small sizes (below 1 ha.) and towards larger sizes (above 4.0 ha.). It means that the socio-economic gaps among the Meghalaya farmers are also widening because farming is the main activity of the rural people and land size variations diversify it. The second social characteristic of landuse is associated with tenancy status of land. It is obvious from the Agricultural Statistics 1970-71 that more than two-third area of the various land holding sizes are being operated by the wholly owned and self-operated tenancy system. Only 17.44 per cent area of the various operational land holding sizes is under operation of rent tenancy system (Table-8.6).

Table- 8.4 : Changes in Cropping Patterns in Meghalaya.

(Area in 000 ha)

Crops	1971-72		1979-80		1987-88	
	Area	%	Area	%	Area	%
Autumn Rice	37.7	-	30.5	-	-	-
Winter Rice	59.9	-	70.1	-	-	-
Summer Rice	0.9	-	1.6	-	-	-
Total Rice	98.5	54.12	102.2	50.34	110.0	50.0
Wheat	0.3	0.16	2.0	0.98	10.0	4.5
Coarse Cereals (Maize & Small Millets)	16.6	9.12	18.4	9.06	20.1	9.1
Pulses	1.4	0.77	2.5	1.23	NA	NA
Oilseeds	5.5	3.02	7.9	3.89	10.0	4.6
Fibres (Cotton, Mesta & Jute)	7.47	13.2	6.50	20.1	9.1	
Other Crops (Including Potato)	14.9	8.19	22.3	10.98	50.2	22.7
Gross Cropped Cereals	182.0	-	203.0	-	220.0	-
Crop Intensity	**110.97**	**-**	**-**	**-**	**115.80**	

N.B. : NA - data not available

Source : Directorate of Economics and Statistics, Ministry of Agriculture & Irrigation (now Co-operation), Govt. of India, New Delhi.

Table-8.5 : Area Under Various Operational Size of Land Holdings (1970-71 and 1980-81).

(Area in ha)

Size of Land	1970-71		1980-81	
Holding	Area	%	Area	%
below - 0.5	8302	3.28	7622	2.57
0.5 - 1.0	29401	11.63	23909	8.08
1.0 - 2.0	78588	31.08	67773	22.90
2.0 - 3.0	67346	26.63	67709	22.88
3.0 - 4.0	30986	12.25	47282	15.97
4.0 - 5.0	18272	7.23	30523	10.31
5.0 - 10.0	16764	6.63	37873	12.80
10.0 - 20.0	3197	1.26	9998	3.38
20.0 - 30.0	-	-	2291	0.77
30.0 - 40.0	-	-	256	0.08
40.0 - 50.0	-	-	725	0.24
50.0 -Above	-	-	-	-
Total	**252856**	**100.00**	**295961**	**100.00**

Source : Agriculture Census 1970-71 and 1980-81, Ministry of Agriculture & Cooperation, Govt. of India, New Delhi.

Table-8.6 : Tenancy Status of Landholdings in Meghalaya (1970-71).

(Area in ha)

Tenancy Status	Area	%
1. Wholly owned & Self-Operated Holding	167393	66.20
2. Partly Owned & Partly Rented Holding	25562	10.11
3. Holdings Partly Taken on Rent	15801	6.25
4. Holdings Wholly Taken on Rent	44100	17.44
Total	**252856**	**100.00**

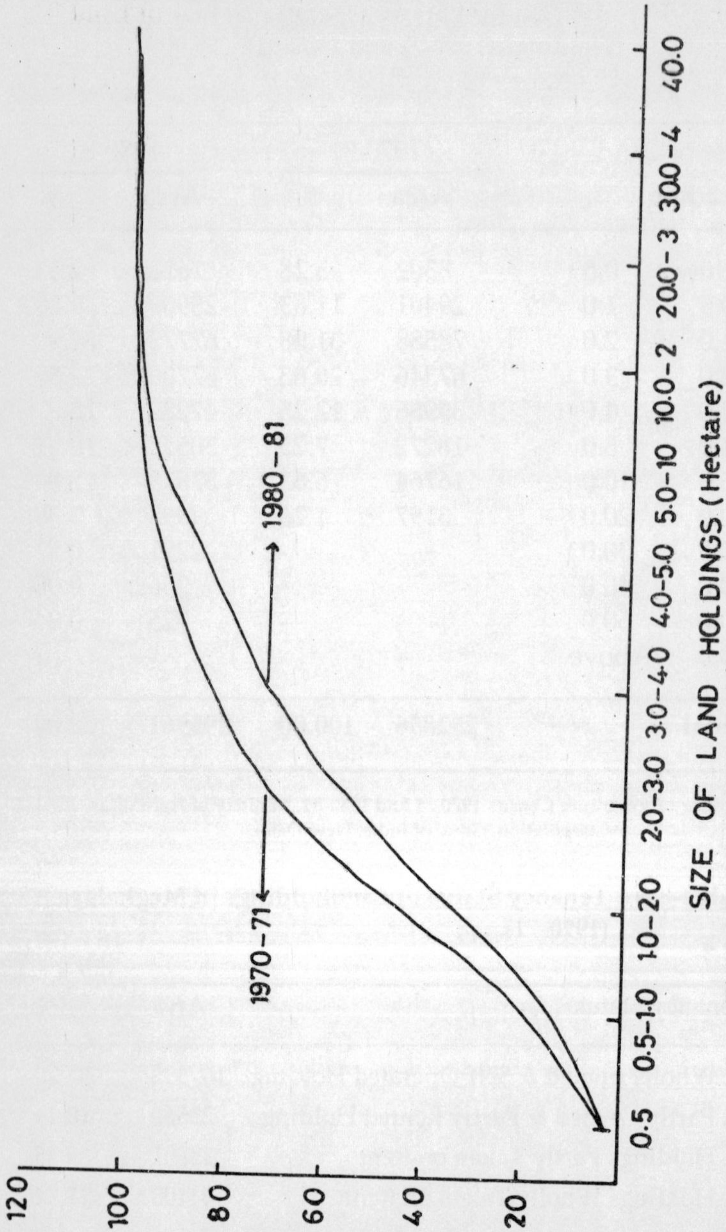

Fig. - 8.3 : Size of Operational Landholdings in Meghalaya.

The rental payment system of these land holdings is traditional; rent is paid in the form of giving share of land products (Table-8.7). On account of traditional tenancy system of land operation, the crop intensity is low. It was only 1110.93 per cent in Meghalaya in 1972-73. The fixed money of land rent payments may be the effective tradition for raising crop yield and farm efficiency of landuse, because the farmers who will occupy land on fixed money rent-lease would use the land in efficient manner.

Table- 8.7 : Terms on Which Holdings are Wholly Taken on Rent (1970-71)

(Area in ha)

Rent Terms	Area	%
1. Fixed Money	2253	5.11
2. Fixed Produce	2343	5.31
3. Share of Produce	6519	14.78
4. Other Terms	32985	74.80
Total	**44100**	**100.00**

In the end, it can be concluded from the above discussion that the landuse of the Meghalaya state is being directly influenced by various physical characteristics of land. But at micro areal processes, the social traditions and economic factors like increasing demands for agricultural products and production prices also play significant role in determining the landuse characteristics.

References

1. Chandra, Bhan, (1988): Spatial Analysis of Potential Soil Erosion Risks in Welo-Region, Ethiopia, *Mountain Research and Development* 1, Vol. 8, No. 2/3, pp. 139-144.

2. Chatterjee, S.P., (1968): La Geographic Regionale De Plateau De Meghalaya', in Singh, R.L. (ed.) *India Regional Studies,* Published for 21st I.G.C., New Delhi, pp. 218-244.

3. Kostrowicky, J., (1972): A Preliminary Attempt at a Typology of World Agriculture,*International Geography* , Vol. 2 (22nd I.G.C., Montreal, Canada), pp. 1097-1100.

4. Rai, R.K. and Panda, P., (1983): 'Hillslope Farming Arround Shillong - A Case Study of Environmental Degradation, in Singh, L.R. and Others (ed.) *Environmenntal Management,* The Allahabad Geographical Society, Dept. of Geog., Univ. of Allahabad, pp. 221-228.

5. Thunen, Von, (1933): Quoted From P. Hall (1966, ed) : *Von Thunen's Isolated State* , Translated by C.M. Warthenburg, Pargamon, Oxford.

6. Vink, A. P. A., (1975) : *Land Use in Advancing Agriculture,* Springer-Verlag, New York, pp. 44-45.

7. Visser, S., (1980) : Technological Change and the Spatial Structure of Agriculture, *Economic Geography,* Vol. 56, pp. 311-319.

8. Visser, S., (1982) : On Agricultural Location Theory, *Geographical Analysis,* Vol. 14, pp. 167-176.

9

Experience of ICAR for Suggesting Improvement to Shifting Cultivation

Arun Varma
ICAR Research Complex for North-Eastern Hill Region, Shillong

---

Shifting cultivation locally known as jhumming, is an agricultural system in which forest land is cleared and cropped for few years and then left for regeneration during the fallow period. Duration of fallow period is much longer than that of cropping period. The types of shifting cultivation are more varied than any other form of landuse system. It is, therefore, useful to consider it in relation to several criteria.

Problems

The problems of shifting cultivation can be classified into four categories, namely, economical, technological, environmental and sociological. Farmer feels that it is landuse system which may belong to past but still works better under his own set of conditions and transitions to more productive landuses, may demand high inputs, advanced scientific knowledge, high capital investment and longer time which he is not in a position to afford.

Jhumia has also become conscious of money-oriented economy, he is no more satisfied with the economic structure in which he is living. His socio-cultural expectations have risen. His attitude towards health, education, culture and shelter is changing very fast. Earlier jhum control schemes were not supported by any technology which can satisfy these expectations. The new alternatives based on experiments conducted in Indian Council of Agricultural Research Complex, Shillong are expected to yield higher returns and maintain pristine eco-system to support higher standard of life.

Approaches

There are two clear cut approaches : (1) the improvement of existing jhumming and (2) replacement of the present form of jhumming by alternative programmes. The present paper focusses attention on the improvement approach of shifting cultivation. Improvement of shifting cultivation will work better in the conditions of the farmers living in remote areas, not accessible to transport net work in their region. Generally speaking, such farmers do not accept alien technology easily.

The important measures which can improve the present form of jhumming are contour bunds between the crops, line sowing, inter-cropping, topo-sequential cropping, use of handy tools in sowing, weeding and harvesting operations and generation of subsidiary source of income through sericulture, mushroom and livestock farming. The following alternatives are based on ICAR researches which would be ideal for shifting cultivation in the hills of North-Eastern parts of the country.

(i) Ideal Landuse : The most significant among the various recommended alternatives is often called as the 'Model Land Use' which can fit in the ideal situations that may not always be available. In model landuse system emphasis is given on minimizing the erosion of soil by giving protective measure of bench terraces on foot hills, on which cropping of food crops is possible, and contour bunds and half moon terraces of mid portion can be utilized for orchard and fruit crop production. Top of the hills should be covered with forests under which sylvipasture can be developed. All the agro-by-products emerging out of the system can be utilized for subsidizing the income of Jhumia by mushroom and live-stock farming. With horticulture crop grown on middle hills, Jhumia is bound to retain long term interest in the land and tied down to settled agricultural system. This system can satisfy the expectations of today's Jhumia.

(ii) Agricultural Landuse : Agricultural landuse is still the most easily accepted occupation of Jhumia. The range of choice under this landuse management is very large. Kharif crops like rice on lower, maize in the middle and tuber crops in the upper regions are most suitable. Maize inter-cropped with soyabean and groundnut is very desirable feature. Rabi crops like mustard, linseed and pea are highly economical. Terrace risers can be utilized for growing fodders. Crop residue, by-products and fodders grown from the terrace risers can be utilized for livestock and mushroom farming.

(iii) Horticulture Landuse : Under the Jhumia settlement schemes of the farmer, an orchard is maintained by the departments and handed over to the farmers in the 5th year of its establishment. It is no longer necessary to convert each and every piece of land of hill slope into terraces for orchard, fruit and vegetable crops. In the North-East, almost all types of fruits and vegetables can be grown. In the plantation crops, the region has been found adequately suitable for tea, coffee and rubber. The dominant role of animal husbandry in the tribal economy also opens a good scope for hortipastoral and sylvipastoral type of landuse.

(iv) Livestock Farming : Livestock as farming on the principle of watershed based landuse system is new concept for the region. But being the self dependent and having the feature for enriching the soil with organic manure it fits both as sole or subsidiary enterprise with all other farming systems. In this landuse, selected hill slope is given protective measure of contour bunds for perennial grass and legume production. Interspaces of contour bunds of gentle slopes is utilized for the production of energy and protein rich concentrate and grain crops. Dairy, piggery and rabbit system can fully sustain on this landuse system, whereas poultry production needs supplementation from outside. Meat and sheep wool can be produced from pasture and sylvipasture systems.

(v) Fish Farming : Fish production also has very high potential in the region. This can be a sole enterprise or part of the farming system. In sole fish production system composite fish culture, cage-culture and running water fish culture are the important systems.

Seasonal fish production systems have also been developed by the ICAR research complex for North-Eastern Hills region, in which dug out cum embankment types of pond can be created. This can be seasonal or perennial water body depending upon the site situations. Such bodies can be utilized for the purpose of irrigation also.

(vi) Agro-based Industry : For the purpose of generating market economy in an underdeveloped area, agro-based industries play very important role. Jhumias produce crops like oranges, lemon, pineapple, tomato, ginger, turmeric, tapioca and other local fruits and vegetables. They fetch very little income. Village-based small processing and packing plants will help the farmer to get far more returns than compared to what they get now. In the same way, rice mills, flour mills and spice mills would also be helpful in rejuvenating village based economy. This system will also generate rural development opportunities. The agro-based industries should be developed by non-conventional energy sources like biogas, solar energy and wind energy implements. This will ease the pressure on other conventional energy sources like diesel, petrol and coal.

Suggestions

Watershed management in jhum control programme involves optimum use of basic resources of land, water and the agro-climate within that landuse. This management of hill slopes calls for scientific survey and investigations. Unlike this no single pattern can be evolved as a common situation for entire region. The programme will have to be devised for each family based on climate, physiography and socio-cultural system of the farmers.

Jhum control is a family programme but it should run along with the biosphere development programme taking village as a unit, because in jhumming farmer was fully dependent upon biosphere to meet his basic needs like fuel, fodder, indigenous medicines and other forest by-products.

Section-C

Geoenvironmental Degradation

10

Environmental Degradation and Its Control

A. Singh
ICAR Research Complex for North-Eastern Hills Region, Shillong

Introduction

Nature has provided enough natural resources for the well being of the people. We human beings are also a part of the resources. Vast changes have taken place in our own life and today most of the villages are facing acute shortages of water, fuel / timber / wood and locally producted food items. The wood lands (forest) are getting thinner and thinner. Wild animals are disappearing. It is not that the villagers do not know of this change but they find themselves in helpless situation in order to meet their immediate needs which is food, shelter and clothing.

During the past two decades, numerous research papers have come out on research and development front but little have been done on the procedural part of development approaches. The result, therefore, is that most of the programmes are taken up in isolated manner. During the 1986, the author had an opportunity to attend a seminar where hundreds of farmers participated. Interesting suggestions were made by the farmers which the author would deal in brief in this paper. We normally take, that traditional farming systems are causing environmental degradation ignoreing the applicability of good systems and methods evolved and used by the farmers in the region. Many of them are unique and conducive to healthy environment. In this paper, attempts have been made to focuss attention on the problems of degradation and possibilities of corrections through peoples participation.

Problems of Environmental Degradation

In the long history of agriculture in the North Eastern Hill Region (NEH), little change has taken place in traditional methods of food production. Most of the systems are followed without soil and water conservation measures and consequently, there is a heavy soil loss in the region (Table - 10.1). Hill slopes are the worst affected areas. Soil erosion has resulted in emergence of flat valleys in between hills which are usually used for intensive paddy production and have become valuable agricultural lands. Sung valley in Meghalaya is one of the rice bowls in the region. Dangerous sign of land degardation has emerged in Meghalaya. In the Jaintia and Khasi Hills, Khlechriat to Amalrem, the Rappleng, the Pynursla and the Cherrapunjee areas; in Garo Hills, the Siju, Mansongirri, Rongram Rongchagiri and Sadengiri areas where stone chips and also parent rocks have cropped up, are severely affected by land degradation because of depletion of soils (Mawthoh 1981).

Excess falling of forests for commercial purpose as well as for shifting cultivation is greatly endangering the environment. This activity is affecting the water resources. Experiences in NEH region indicate that many major irrigation and drinking water schemes are failing because the removal of the trees from forest areas.

Table- 10.1 : Soil Erosion from Traditional and Forest Landuse Systems.

Landuse	Soil Erosion (tonnes/ha/year)	Remarks
Forest	0.04 to 0.5	Estimation at site indicates that soil erosion from shifting cultivation
'Bun' method of cultivation	40.0 to 50.0	areas are as high as 5.1 to 170.2 tonnes/ha/year.
Jhumming	5.1 to 83.3	
Pineapple along the slope	24.0 to 62.6	

Awareness on Environmental Degradation

People know about the changes taking place in their surroundings and environment. Data reproduced in Table - 10.2 from a recent study in Garo Hills strongly supports this view (Rao *et al.* 1989).

In the process of evolution, several good agricultural systems have emerged which suppots the population entirely on the local and renewable resources. Such systems have strong bias towards the environment and are still in the process of development. Improved methods of shifting cultivation in Nagaland (using elder tree for fertility build up), Apatani plateau in Arunachal Pradesh for valley land management and bamboo drip irrigation system in Meghalaya reflect the ingenuity of the local people on environmental issues. The problem is, therefore, not much of creating awareness but is of thinking as to how to combat through the development processes which are in vogue.

Table - 10.2 : Extent of Awareness about Changes in Environment amongst Shifting Cultivation from Garo Hills.

Changes in	Responses from 44 Randomly Selected Farmers (%)		
	Yes	Do not Know	No
Rainfall Pattern	84.09	9.09	6.82
Temperature	81.82	18.18	-
Drinking Water	70.45	22.73	6.82
Availability Soil Fertility Status	81.82	6.82	11.36
Availability of Forest Product of Commercial value	75.00	-	25.00
Availability of Fuel Wood	97.73	-	2.27
Resence of Wild Animals	100.00	-	-
Availability of good Quality Housing Material	100.00	-	-

An Experience on Awareness : People's participation in development programmes is a must. Experiences so far indicate poor responses on this aspect. In one of the open house seminar held at Aizawl (Mizoram) during 1986, this question was examined in depth in consultation with farmers who were attending the seminar. Their suggestion on the isuue of peoples participation was as follows :

1. Their (farmers) need should be decided by themselves.
2. All development departments should assist the farmers at a time instead of approaching individually at different times.

3. At the time of execution, there should be only one agency responsible for assisting the farmers.
4. There should be one window at district level where farmers can approach for solving their problems.
5. There should be periodical review of the programmes being implemented with farmers, representatives participation.

If we examine the basic philosophy behind water-shed management programmes, one would see the above demands/suggestions are nothing different than the concept on which development programmes in agriculture has to be undertaken. The concept provide overall development and management through peoples participation. Plans for individual farmers/land users and management of resource through watershed authority (which is none except the village society elected for the purpose) are the essential requirement envisaged in watershed management programmes.

Technology for Conservation and Management of Resources

ICAR Research Complex, Shillong in its research programmes have been conscious from the very begining particularly in working out alternatives to traditional farming systems. Conservation based farming/landuse systems have been studied. These studies have revealed the feasibility of using local natural resources, *viz..*, man and material for checking soil erosion/degradation while practising desired landuse(s). Rainfall retention in situ and soil protection can be achieved similar to forest landuse while producing food, fruit and water crops. Contour bunds, bench terraces, half moon terraces, grassed waterways, dugout cum embankment type of water harvesting ponds are the major conservation measures. Micro watersheds as a base for landuse/ farming systems is only an additional requirement in the technique which helps in providing environment security to alternatives. Soil losses and rainfall retention from such systems are presented in Table - 10.3.

Dugout-cum-embankment type of water storage structures with the kind of landuse technique experimented and suggested by ICAR Complex receive 75.0 and 22.5 per cent of water as base-flow, surface flow and rain respectively.

Table- 10.3 : Soil Erosion and Rainfall Retention in Scientifically Managed Landuse/Farming Systems.

Landuse	Soil Erosion (tonnes / ha / year)	Rainfall Retention (% of annual total)	Remarks
1. Agriculture (Food Crops)	0.00 - 7.70	95.03-100.00	Soil erosion control measure are developed with the use of locally available materials
2. Agriculture (Fodder)	0.00 - 0.16	98.68-100.00	
3. Horticulture	Trace- 4.37	96.10- 99.75	
4. Agri-Horti-Silvi-Pasture	Trace- 1.22	98.27- 99.99	

Measures to Control Environmental Degradation

Scientific land management at individual land users level and management of natural resources of the village/community in an organized manner may be only solution for controlling further degradation of the environment. It calls for organizing village resource management society, scientific landuse plan for each and every individual farmer, development plan for the villagers and organizing social welfare activity for villages which would go a long way in developing towards self reliance for sustained and ecologically sound food production system.

References

1. Mawthoh, P.R. (1981) : Constraints in Jhum Control and Soil Conservation Activities in Meghalaya, *Proceedings of Seminar on Socio-economic aspects in Soil Conservation activities in Meghalaya with special reference to North Eastern Region,* Soil Conservation Deptartment, Government of Meghalaya, India, 63-69.

2. Rao, N. V., Reddy, B. S. N. and Singh, A. (1989) : Altitudes and Awareness of Jhumias : Issues in Eco-development, *Journal of Institute of Social Research.,* 9(1 & 2), 1-5.

11

Geomorphic Processes and Environmental Degradation

R. K. Rai
Department of Geography, North-Eastern Hill University, Shilllong

The environmental crises of which many voices warn us is an element of many dimensions. Environmental degradation presents itself on an areal scale running from local to global and on time scale from the present to the indeterminable future point of time. The threat of crisis ranges in magnitude from deterioration of small ecosystem to the total destruction of the biosphere.Today countries all over the world and especially developing countries are affected by the impact of development and resultant environmental crisis, which are manifested in pollution of air, water and land, rapid deforestation and depletion of non-renewable resources, soil erosion, loss of top soils. Development and the environment are inter-twind issues. All developmental activities put extra demands on our natural resources and affect the environment. Land, water and air are the most important natural resources which support all living forms. Unjust economic relations (whether at international, national or local levels) lead to ecological disturbances, resource depletion and environmental degradation. These environmental crises in turn have a disastrous affect on the development of developing societies. Environmental degradation is inevitable under the operation of natural geomorphic processes. The resultant outcome is an equilibrium state at given point of time. However, the problem becomes grave or the imbalances come into play or the equilibrium is disturbed when man interferes with the natural processes and accelerates these processes.

The chief geomorphic processes are :

1. Weathering,
2. Mass Wasting,
3. Erosional Processes, and
4. Soil Formation.

In the natural course of time, any terrain due to the impact of these processes tends to evolve landforms in a cycle, which is known as 'cycle of erosion'. But the greed of man to harness natural resources most often than not over exploits the land resources. This activity is mainly responsible for environmental degradation.

Weathering is the initial phase in the denudational history of the landscape. The surface rocks are weathered and the loose material is transported by running water. The process of erosion is very limited unless the agents of transport are carrying a load of debris.

Only a few centimetres of the soil of the land formed over millions of years due to natural denudation of rocks, is used for agriculture. But over use of land, deforestation and other mismanagement result in rapid soil erosion. It is estimated that world over in the next 20 years, an area equal to the entire cultivated area of India today (that is 140 million hectares) will be lost if soil erosion continued unabated. Nearly 5 to 7 million hectares of good agricultural land are being eroded every year all over the world. The other causes of environmental degradation of land are water-logging and salinity. As many as 175 million hectares of land constituting about 53 per cent of India's total area are subject to serious environmental stress due to soil erosion, water logging and salinity. The total annual soil loss in India can be valued as Rs. 700 crores in terms of plant nutrient loss.

Mass wasting is an important gravity controlled movement of materials over the earth's surface. This evolves slow to sudden downward and outward movement of material, and perhaps is the most important agency that sculpture and reduce the earth's surface. Landslides are one of the major classes of mass wasting. Landslides are of profound interest to the earth's scientists, environmentalists and engineers. Landslide phenomenon is one of the most widespread and effective agent sculpturing the earth's surface. A landslide unforeseen and improperly provided for, may destroy settlements or engineering structures like road, bridges, buildings, etc. or impair their usefulness. It may also bring sudden deaths to people who have trusted these structures.

Landslides denote downward and outward movement of slope forming materials, primarily composed of natural rocks, soils, artificial fills or combination of these materials. They are more frequent where the developmental (constructional) activities have modified slope profiles, particularly making the profile segments steeper in sections. This makes the slopes unstable with a tendency to fall as and when the equilibrium is disturbed. Incidentally, all the active landslides in Meghalaya are encountered along National Highway No. 40.

Fluvial process is the most dominant geomorphic agent in sculpturing the landscape of the area. The activities of streams or running water could be either erosional or depositional. Progressive dissection and degradation of the higher lands are caused by fluvial processes. Due to erosion, different types of valleys and interfluvial ridges are formed depending on the stage of development and the lithology of the river and structure of the area.

Geomorphological Parameters

Within the contemporary concern for environmental management, many problems relate to the interaction between man, land and water. The study of natural environment frequently recognises that when man uses land, he has to accommodate its relief, materials and water resources to his purpose. The environmental management calls for understanding the geomorphology of an area before the land is used as a resource. In access, where developmental activities underestimate or ignore geomorphological characteristics, it is bound to create serious environmental imbalances.The slope characteristics, the relief of an area, the physio-chemical characteristics of soil, the hydro-geological conditions and subsurface geology are some of the important geomorphological parameters. The same parameters determine the suitability of a location for construction of dams, railways, highways, electrical powerlines, industries, townships, irrigation distribution channel net work and defence strategic installations.

Being exquisite in terms of beauty, this area possesses a huge wealth of natural resources such as forests, minerals, building materials, wild lives, medicinal plants, etc. Due to unscientific and over exploitation of these natural resources and the developmental activities, the environment is deteriorating and hence causing imbalances in ecosystem. Thus, this area is experiencing danger from the acceleration of unplanned developmental processes and exploitations.

This area, as a whole, is extremely susceptible to large scale soil erosion. Shifting cultivation on steep slopes are proving disastrous and resultant soil erosion, many areas have turned into barren hills and have become unfit for any economic agricultural use. With the increase in population and declining productive potential of land, more and more areas have been brought under cultivation and period of fallow has decreased gradually. Cutting the trees for timber, fuel and fodder, road construction, hydro-electric power projects and other developmental projects have resulted into reduced forest area. Most of the environmental ills in the country are directly or indirectly related to deforestation. Conservation of forest wealth is, therefore, most vital for our survival. It has resulted quicker surface run-off in rainy season with ensuing soil erosion, gully development and landslide.

To highlight the effects of damage due to deforestation, an example around Barapani has been cited here. In an area of 1935 sq. kms. shows the changes in landuse pattern (Table - 11.1).

Table- 11.1 : Landuse Changes in the Barapani Area of Meghalaya (between 1910-11 & 1966-67).

Landuse	1910-11 Area (Sq. Km.)		1966-67 Area (Sq. Km.)	
	Total	%	Total	%
1. Forest	38.5	26.9	35.5	24.8
2. Agriculture	11.3	7.9	-	-
3. Shillong Township	4.3	3.0	9.9	6.9
4. Deforested	77.9	54.5	87.6	61.3
5. Completely Deforested	11.0	7.7	-	-
6. Lake	-	-	10.0	7.0

It is worth mentioning that in this area, the primary forest has almost disappeared. The present forest cover represents secondary pine forests only. This secondary pine forest are also under indiscriminate exploitation. As a result, large patches of deforested land is visible on the slopes and forests older than 25-30 years are not seen.

Another important factor which has endangered for the environment is the construction of roads. Though, these constructions are very essential for the development, but necessary measures and precautions are seldom taken and hence they are also treated as threat for environmental degradation. Many of these roads are unscientifically planned thus causing hillside slope and producing a huge amount of debris. The debris is allowed to slide down the slopes into the valleys and, in turn, carried away by streams, leading to rapid siltation. Moreover, the sliding of the debris destroys plants completely. Indiscriminate construction of roads, bridges, heavy building and other developmental works like canal, dam etc. must be discouraged. Such activities should be recommended only after a thorough geological analysis.

The landslides although can not be prevented or can be avoided if suitable measures are taken before a developmental scheme is implemented. They are as given below :

1. Proper guard maintenance of slope.
2. The slope cut faces should be rounded rather than making them sharp.
3. Breast and retaining walls to be made.
4. Benching and terracing of slopes.
5. Proper drainage galleries and drains to be provided.
6. Wire mesh to be provided in rock fall areas.
7. Slope profiles to be suitably modified in slide prone areas.
8. Vegetation cover along the highway areas not to be disturbed.
9. Total ban on agriculture, grazing and additional residential construction activities along both sides of highways.

Considering serious nature of the consequences in the fragile ecosystem of the area, there is need of identifying the geo-dynamically vulnerable area on the basis of geological structures, lithology, nature of topography and vegetal cover for the better environmental management.

There is a need to re-establish the harmonious relationship between man and nature. The main objective of forest management should be soil and water conservation alongwith other natural resources like timber, vegetal cover, wild life etc. Deforestation should be stopped and afforestation must be resumed.

Environmental awareness should be a part of education to achieve wider participation of local people so that a continuous flow of technically skilled personnel would be available at village level for extension activities. Jhumming should be stopped and alternatives should be sought out. In reviewing the overall situation, a very systematic appraisal of soil, minerals, forests and water is warranted. The environment must be preserved. Deforestation should be stopped and afforestation should be encouraged. The national environmental awareness campaign should be taken up in this area on a large scale which would help in increasing environmental consciousness among the people of this area.

The non formal methods of environmental education can play a crucial role in developing an awareness and consciousness among the masses.

12

Management of Hill Resources

R. N. Prasad
ICAR Research Complex for North-Eastern Hill Region, Shillong.

Agriculture is the backbone of the Indian Economy. About 70 per cent of the country's present population of 815 million is dependent on agriculture and contributes nearly 30 per cent of the Gross National Product (GNP). The population of India is increasing at present at the rate of 2.26 per cent annually and has risen from 238 million in 1901 to 361 million in 1951 and 820 million in 1989. The population is expected to touch the figure of 1000 million by 2000 AD. The country supports about one-sixth of the country's population on only 2 per cent land area. The size of operational holdings is also decreasing. On an average, 72.6 per cent of the farmers own holding is less than 2 ha., 24.4 per cent between 2 and 10 ha. and only 3 per cent over 10 ha. The per capita availability of arable land has been declined from 0.36 ha. in 1951 to 0.25 ha. in 1982 and is expected to decline further to 0.15 ha. by 2000 AD as against the required of 0.4 ha.

With the rapid development of viable technologies in agriculture after independence, country has made commendable progress in the total food grain production from 51 million tonnes in 1950-51 to 172 million tonnes in 1988-89. Per capita availability of food grain improved steadily from 153 kg. per annum (1950-55) to 175 kg. per annum (1984-89). Based on the consumption pattern, it is felt that the required per capita availability of food grains by 1995 should be 195 kg. per year. Thus, we need to enhance per capita availability of food grain production at a much faster pace through concerted effort in a systematic manner to a level of 210 million by 1995 and 250 million tonne by 2000 AD.

Despite endowment of good soil, abundant sunshine, highest percentage of land under cultivation, ample water resources etc. in India, the average crop productivity per unit area is one of the lowest in the world. This dismal situation is mainly attributed to the defective utilization of the various resources such as, soil and water and low use efficiency of the inputs.

In the recent past, there has been a growing national and international concern for conservation of soil as a vital life support system. There are historical quotations which remind us that some civilizations perished as they did not take adequate care of their soils. As a natural habitat of fauna and flora, soil is a basic resource for all our agricultural activities. India, indeed, had been very fortunate to be blessed with one of the richest endowments in this respect. Therefor, our farming ancestors were able to make agriculture a developed art much ahead of other people and civilizations. However, over the time, as biotic pressures built up beyond the carrying capacity of this resource, these contributed to the misuse or overuse of crop lands, over-grazing and over-stocking of rangelands, over-exploitation of forest lands for fuel, timber, mining as well as industrialization. On account of the disturbed harmony of our natural environment, we have been witnessing increasing instances of soil resource degradation, in the form of sheet and rill erosion, land denudation, sand dune formation, shifting cultivation, mine soils and pollution of soils with industrial wastes. At some places, degradations have reached disastrous proportions, and if not corrected in time, will have a devastating effect on our future population.

We have a responsibility to conserve this resource for the coming generations not only in an unimpaired state but with improved productivity. This demands monitoring of our efforts in this direction and development of management practices which would help to restore the productivity of our soils.

Soil Resource Problems

Out of the total geographical area of 329 million hectares, about 265 million hectares represent varying degrees of potential for biological production which include 143 million hectares of cultivated land, 82 million hectares of forests and pasture lands, and about 40 million hectares of culturable waste and fallow lands. About 174 million hectares of this biologically productive land suffers from soil erosion, gully and ravine formation, salinity, waterlogging, shifting cultivation etc. Besides these problems, the cultivated areas also suffer from water-stress, fertility depletion and chemical toxicity.

(A) Land Degradation

In the Himalayan region, the important factors of land degradation are prevalence of shifting, bun cultivation and deforestation etc.

Shifting Cultivation : Shifting cultivation is being practised by 492000 tribal farm families of this region affecting 2.7 million ha. of sloppy land even beyond 100 per cent slopes. Total area affected by this system in the country is 4.36 million ha. Biotic interference in jungle cutting, burning, clearing and dibbling of seeds accounts for a considerable amount of loose soil material, ashes, earthworm casting and detached soil/stones to roll down the foot hills causing soil erosion to the tune of 3.7 t/ha.

(B) Soil Erosion

Soil erosion problem in shifting cultivation is mainly of splash and wash. The soil erosion from hill slopes (60-70 per cent) under first year, second year, abandoned jhum and bamboo forest was estimated to be 147 and 170 t/ha./year in first year and 30 and 8 t/ha./year for the second year respectively. Second year shifting cultivation was found to be more hazardous than the first year. Nature takes almost about 500 years to make 2.5 cm of soil. Due to increase in the pressure on land, the jhum cycle has reduced to 4-5 years in Meghalaya, 5-10 years in Mizoram and Tripura, 6-15 years in Nagaland and Manipur and 1-17 years in Arunachal Pradesh, giving practically no chance for natural regeneration of soil fertility. Shifting cultivation prefers bamboo's secondary growth to a primary or secondary tree forest as it is easier to cut, dries rapidly and burns well, yielding plenty of K-rice ashes for successful jhum crop. But with shorter jhum cycle, weeds like *Eupatorium* spreads widely. It tends to check the growth of coarse grasses like *Imperata Cylindrinca,* but once *Imperata* invades a badly degraded land, restoration of forest is time consuming and expensive. Cultivation of tuber, *rhizomatus* crops on raised bag along the slope 'bun method' causes soil erosion to the tune of 40-50 t/ha/year while pineapple along slopes removes 24.0 to 62.6 t/ha/year. The entire Shillong plateau has been denuded by this method of cultivation. It is estimated that 181 million tonnes soil is lost every year from this region due to the above defective system of farming.

(C) Deforestation

The reckless felling of forests for timber, fuel and pulp industries in the catchment areas of the rivers is responsible for serious soil erosion by water. As a result, there has been decline in area under forest in all the states except Sikkim and Arunachal Pradesh. Forests are being denuded in the country at the rate of 1.5 million ha/year. It is reported that the forest cover of over 60 per cent of land area in Himalayas has been reduced to about 25 per cent. Recurring floods in Gangetic plains and Brahmaputra valley are the result of deforestation.

According to the National Forest Policy, one-third of the total geographical area in India should be under forest including 60 per cent in the hilly region and 20 per cent in the plains. Though figures of 67.4 million ha. (22.2%) has been reported by the Forest Deptartment, the actual area under forests with more than 30 per cent density is hardly 33 million ha. (10%). It is estimated that the total area under forest with the density of more than 10 per cent has decreased from 55.18 million ha. to 47.08 million ha. during the period 1975 to 1982.

(D) Loss of Genetic Resources

Environmental destruction caused by shifting cultivation and deforestation are destroying many wild varieties of even domesticated crop plants, trees and animals. This region is considered one of the centre of origin of rice and citrus and secondary origin of maize. About 7000 endemic species are found in India, of which 3000 alone accounts for in the Himalayas and Khasi Hills. Preservation of these are essential for future exploitation. So far over 8000 germplasm of rice and 17 of citrus have been collected from this region.

(E) Water Resource

Out of a total volume of precipitation of 400 million hactare metre (m. ha. m.), 70 m. ha. m. are reported to be lost through evaporation, 180 m. ha. m. go as run-off and 150 m. ha. m. enters into the soil. Out of 150 m. ha. m., 100 m. ha. m. are actually retained as soil moisture, while about 40 m. ha. m. percolate down to recharge the groundwater aquifers. Out of 180 m. ha. m. of run-off, about 20 m. ha. m is stored in reservoirs, tanks, ponds etc. whereas 160 m. ha. m. goes back to sea.

There is ample of water resources in NEH Region. Almost 10 per cent of the total rain of the country is received in this region. Brahmaputra Basin including Barak have an average annual flow of 54.0 m. ha. m., however, utilisable flow is only 2.4 m. ha. m. Only limited area is under irrigation in this region. In Nagaland, the streams have diverted to lower terraces of irrigation. Shifting cultivation and deforestation have affected badly the retention of water in situ. As a result, major portion of the rain water is carried as run-off resulting into floods in the plains. Siltation of reservoir and scarcity of drinking water during lean period even in the hills are not uncommon.

(F) Grazing Resources

The livestock population in India is nearly 15 per cent of the total livestock, but produces only 3 per cent milk, 1 per cent pig produce, 0.14 per cent pork, 4 per cent sheep produce, 1.3 per cent wool. The projection of green and dry fodder has been estimated at 992 and 832 million tonnes by 1990 and 1136 and 949 million tonnes by 2000 AD. The present feed and fodder resources of the country can meet only 46.6 per cent of the requirement.

Forest grazing resources are also dwindling as we are losing roughly 1.5 million ha. of forest every year. The 67.33 million ha. of forests cover are important source of forage and grazing. The number of animals in the forests of Jammu and Kashmir is about 1.19 million, out of which, 1.16 million are goats and sheep. In U. P. about 2 million animals grazed in forests, of which cattles, buffaloes and goats and sheep constituted 1.22, 0.25 and 0.7 million respectively.

Permanent pastures accounts for 12 million ha. (3.94% of geographical area) and form the main grazing sources. North Eastern Region has great potential for pasture development. The density of livestock population per hectare of geographical area basis is the highest in Assam (4.2), followed by Tripura (3.19), Meghalaya and Sikkim (1.21) and

Manipur and Nagaland (0.69). Central plateau of Khasi Hills can be developed into pasture land for goats.

(G) Soil Fertility Depletion

The complexity and magnitude of soil fertility research have assumed much greater dimensions these days due to rapid change from subsistence agriculture to dynamic commercial farming and also due to adoption of multiple and intensive cropping for increasing crop production, short duration, fertilizer responsive and photo-intensive varieties of important crops. This is causing faster and greater nutrient depletion from soil. Removal of top fertile soils causes further nutrient imbalance in soils. About 6000 million tonnes of fertile top soil and 8.4 million tonnes of plant nutrients are estimated to be lost by soil erosion annually.

Severe reduction in soil fertility is bound to occur. If suitable steps are not taken simultaneously to restore it through application of balanced inorganic fertilizers in conjunction with organic manure.

Basic Issues

Vital issues are the scientific management of our natural resources such as soil, water, plant, man, animal on watershed. It should be based on integrated manner for optimization of production on sustained basis by judicious practices of soil and water conservation for the prosperity of posterity.

Research Efforts

ICAR Research Complex for North-Eastern Hills Region, Shillong has been working on integrated farming systems such as, livestock, agriculture, agri-horti-silvi-pasture, horticulture, agro-forestry, forestry etc. and useful informations have emerged.

1. 90-95 per cent of rainfall can be retained within the micro-watershed by adopting soil conservation mea-

sures such as, contour bunding, bench terracing, vegetative bund, half moon terracing.

2. There was practically no soil loss.
3. Agriculture alone is not economically viable unless it is integrated with livestock/poultry.
4. Agri-horti-silvi pasture system was most suitable for maximizing productivity on sustained basis.
5. Intercropping of ginger, turmeric, groundnut and soyabean with mandarin orange/guava was economically viable.
6. Terrace riser can be utilized for raising fodder grasses and legume.
7. Excess biomass can be utilized for growing mushroom.

How To Implement

Scientifically, it has been proved that for integrated development, watershed or catchment of a stream/ river is appropriate area for use and management of natural resources. Since, man is the most important single factor, who influence the entire network, it is being considered that a village may be taken as a unit for development. The concept of watershed or catchment can well be adhered to through micro-watershed based landuse systems.

(A) Village Water Management Society

Effective management and exploitation of natural resources calls for organizing a 'Village Resource Management Society'. All the families residing in the area needs to be the member of the society. The major function of this organization will be to help in planning the master landuse plan of the village, organizing input, supplies, marketing infrastructure, social activities, generation of village fund, employment opportunities etc.

(B) Individual Land Use Plan :

Each and every farmer should have individual landuse plan depending on his interest and resources. While planning for landuses the farmers should be assisted rather than being suggested with new ideas after having been exposed to new ideas, nothing should be insisted or forced in the plan. Research experiences indicate that at individual level the land can well be managed with the use of local natural resources in the following manner.

(i) Master Plan of the Village : After having developed landuse plans, the left out area common to the community can be planned to meet the community needs (pasture, water supply, fuel needs etc.) and resource generation. Specific commercial forest plantation for timber, fuel, fishery development, plantation crops and forest resources for wild animals etc. should be planned. Combining individual plans with the common plans, the master plan canbe prepared.

(ii) One Time Investment : Concepts such as, one time investment for village development similar to any industry is now very much in the talk. If resources are modified, developed and used, it should not be difficult to generate fund for subsequent needs. It has been successfully experimented in some case studies. The recent concept of Government of India to channelise funds to villages through Panchayats can well be adopted in Meghalaya through the existing village machinery which is quite favorable for organizing Village Resource Management Societies. Any investment/development in the village should be routed through the Village Resource Management Society.

Development - A Permanent affair

One of the serious defect in present development programmes is the lack of follow up action or maintenance. The follow up programme in development works related to environment is must. It can well be taken through village resource development society. Even if individual development departments work in isolation, it would be feasible to absorb the programme in the overall integrated concept through the village society.

Section-D

Forest and Wildlife

13

Wildlife Preservation And Community Action

Vincent T. Darlong
Ministry of Environment and Forests, Government of India, Regional Office, Shillong

Introduction

Environment is now part of the current style of almost every public deliberations - locally, nationally or internationally. Environment is the sum of all the external conditions in which life exists. A very important component of the environment is 'wildlife'. Themes on wildlife conservation / preservation are now familiar topics. And yet, there depletions continue to be alarming inspite of governmental laws and regulations to conserve and preserve them. Wildlife encompasses, in a broad sense, all the living plants and animals, found in nature in their wild state. More and more people are now showing increased desire to acquire greater knowledge about nature and wildlife. In the present context, the scope of wildlife is restricted to the wild animals and birds.

We may recall that from the dawn of civilization, wildlife played vital role in the development in the human culture and heritage - as source of food, fuel, tools, clothings, folklores, religious beliefs and so on. With the transformation of human societies from food-gathering to food-processing (agriculture) some species of wildlife were domesticated to serve man directly, while others continued to be hunted for their flesh, hides, fur, ivory, horns, beaks and for human pleasure.

Population explosion in the last few decades, demanding development in various spheres, has resulted directly or indirectly in sudden and often far reaching disturbances in several natural ecosystems. Rapid urbanization and industrialization, construction of dams and hydro electric projects, buildings and roads, agricultural expansion and mining operations, are but few examples of agents bringing changes and disturbances in our natural ecosystems. The chain reactions of these are deterioration of environmental quality and depletion of wildlife resources.

In our context of Meghalaya, a naturalist who might have been in this part of the country, say about one hundred years back or even lesser, would perhaps shudder to see the changes if he would visit the same areas now. Therefore, at this juncture, it would be worthwhile to take stock of what we still have in relation to our forests and wildlife, and discuss some possible areas of community action and participation to create awareness and enhance the practices and efforts of wildlife conservation and preservation. General constraints to such efforts with particular reference to this part of the country are also briefly touched upon as an aid to generation of new ideas and avenues towards solution of such problems.

Forests and Wildlife in Meghalaya

To an industrialist, forest is the source of raw materials; to a forest department and other agencies, it may be looked at as the source of revenue; to rural villagers it is the basis of economy, social and cultural life. Similarly, vegetal cover or forests to wild animals and birds mean food, shelter, propagation and survival.

Table-13.1 tells us the extent of government controlled forests in Meghalaya as Reserved and Wildlife Sanctuaries. Statistics indicate that about 38 percent of total geographical area of Meghalaya is forest area in one form or the other. It further reveals that, of the forest areas, government controlled is nearly 11.65 per cent, which is only 4.43 per cent of the total geographical area. The rest of the forests are under the control of District Councils, Communities, Villagers or Private. With this unique situation, it is here that the importance of community efforts and actions far out weigh governmental efforts in nature conservation/preservation.

Scientists estimate that upto 30 million biological species inhabit our planet. According to zoologists, India alone has about 500 species under atleast 135 genera of mammals, 1230 species of birds, 220 species of snakes, 150 species of lizards, 30 species of turtles, 3 species of crocodiles, 142 species of amphibians, 105 genera of fishes, 30000 and odd species of insects, besides innumerable other invertebrates.

Zoo-geographically, Meghalaya falls within the 'Assam Zone' of the faunal distribution. Within the 'Assam Zone' it is reported that there are about 85 genera of mammals spread over 148 species and 186 subspecies. Some of the animals and birds found in the state are listed in Table-13.2. Several new vertebrate species were described from Meghalaya (Table - 13.3), beside numerous invertebrates like coleoptera, orthoptera, hymenoptera, other arthropodes, namatodes, etc. not included in the present list. Today many of these animals find safe refuge in the Balphakram National Park, Nokrek Biosphere Reserve, Siju Wildlife Sanctuary, Nongkhyllum Wildlife Sanctuary and perhaps in other reserved forests. These animals are also available in the areas controlled by District Councils, forests belonging to the community or private individuals.We don't have data to illustrate how many of these animals have already been lost from such forests. About 100 years ago, Rhinoceros were reported to be abundantly present in Garo Hills adjoining Goalpara areas.

Table-1 3.1: Forests and Sanctuaries under the Control of Government in Meghalaya.

Districts	Reserved Forests / Sanctuaries etc.	Area in ha.
West Garo Hills	Dibru Hills RF	1502.195
	Baghmara RF	4428.886
	Angaratoli RF	3010.865
	Rewak RF	647.497
	Emangiri	828.797
	Tura Peak Catchment RF	418.631
	Balphakram NP	22000.000
	Nokrek BR	4748.000
	Siju WLS	518.000
	Baghmara Pitcher Plant Sanctuary	2.700
East Garo Hills	Chimabangshi RF	2328.484
	Dhima RF	2071.993
	Dilma RF	258.999
	Rajasimla RF	1812.994
	Ildek RF	259.999
	Darugiri RF	1035.996
	Rongrengiri RF	3625.988
	Dambu RF	1812.994
	Songsak RF	2330.992
East Khasi Hills	Riatkhwan RF	391.331
	Nongkhyllem RF	9690.841
	Umsaw RF	43.706
	Shyrwat RF	44.160
	Riat Laban RF	204.660
	Nongkhyllem WLS	29000.000
	Upper Shillong PF	766.474
	Short Round PF	113.312
	Laitkor PF	324.920
	Umkhuti PF	13.640
	Green Block No. 2 PF	20.749
Jaintia Hills	Saipung RF	15035.308
	Narpuh (Block I) RF	6241.880
	Narpuh (Block II) RF	9867.870

Abbreviations : RF=ReservedForest;BR=Bio-sphere Reserved;NP=National Park; WLS = Wildlife Sanctuary; PF = Protected Forest

Table-13.2 : Common Mammals and Birds of Meghalaya with their Vernacular Names.

Common Names	Khasi Names	Garo Names
(A) Mammals		
Elephant	Hati	Mongma
Tiger	Khla	Matcha
Leopard	Khla Basa	Matchapeng
Indian Bison (Gaur)	Masi Prut	
Wild Buffalo	Mynthra	Matma Burung
Golden Cat		
Capped Langoor		
Hoolock Gibbon	Huleng	Huro
Leopard Cat	Ksar	Mengo Burung
Himalayan Black Bear	Dngiem	Mapil
Serow	Khiat	Matrong
Wild Dog (Dhole)		Achak Burung
Sambar	Sier	Matchok
Barking Deer	Skei	Balgitchak
Slow Loris		
Flying Squirrel		Matwan
Wild Boar	Sniang Khlaw	Wak Burung
Others	Ksih	Matram
(B) Birds		
Hornbill	Kohkarang	Renggok
Duck	Hankhlaw	Dogep
Peacock Pheasant		
Golden Oriole		
Hawks	Kbeit	Doreng
Kaleej Pheasant	Syiar Tung	Dogrik
Jungle Fowl	Syiar Khlaw	Domesal
Hoopoo	Sim	Sohphoh
Wood Pecker	Sim Puhdieng	Dotileng
Emerald Dove		Dokru
Falcon		
Owls	Dkhoh	Dopo
Partridge		
Quail	Syiar Khiar	
Hill Myna	Myna	Dosing/Myna

Table-13.3: New Animal Species/Sub-Species Described From Meghalaya.

(A) Reptilia

Takydromus Sexilineatus Khasiensis (Blgr.)

(B) Amphibia

Rana garoensis Boul.
Rana danieli Pillai
Rana mawphlangensis Pillai
Philautus Garo Boul.
Philautus shillongensis Pillai and Chanda.
Philautus cherrapunjiae Roonwall and Kripalani
Mycrophyla berdmorei (Blyth).
Bufoides meghalayana (Yazdani and Chanda)
Pedostibes kempi (Boul.)
Leptobrachium hasselti Tschudi.
Hyla annectens (Jerdon)

(C) Pisces

Danio meghalayensis.
Danio deyi.
Puntius shalynius.
Mesonoemacheilus reticulofasciatus.
Pillaia indica.
Garo khajuriai.

Elephants Distribution in Meghalaya

Elephant depredation and lakhs of rupees paid as compensation to affected villagers every year is well known feature in Meghalaya. Perhaps this is a way of protest by these large but meek animals as a result of disturbance and shrinkage of their habitats and migratory routes.

The total number of elephants in Meghalaya is estimated to be 2500-3000 individuals or so. Accurate estimate is difficult as interstate movements are frequent among these animals, particularly those from Khasi and Jaintia Hills with the neighboring Assam. Nearly 80 per cent of the estimated population is reported to be distributed in Garo Hills (East and West Districts) mainly in four disjointed concentration belts:

(1) Balphakram National Park and its adjoining areas lying specially in between river Sameswari and Maheskhola,

(2) Songsak - Rongrengiri Reserved Forests and its surroundings,

(3) Nokrek - Arabela area (this area has now been declared as Biosphere Reserve) and

(4) Dadenggiri - Rongrengiri - Phulbari area.

About 12 per cent is estimated to be distributed in the East and West Khasi Hills Districts in two major adjointed belts, namely,

(1) Nongstoin - Sonapahar area touching Assam borders and

(2) Nongkhyllum - Jirang - Sonapur zone touching Assam borders.

The remaining 8 per cent of the population is estimated to be present in the Narpuh and Saipung Reserved Forests of Jaintia Hills. In the past, elephants from Kaziranga were known to have come to these areas via Karbi - Anglong forests.

Special Problems of Wildlife Protection

(1) Prolonged and Heavy Monsoon : This part of the country experiences a prolonged rainy season with usually heavy rainfall. In many areas, devastating floods, landslides and other side effects of deforestation contribute and accelerate loss of habitats, as well as lives of the animals.

(2) Difficult Terrains and Inadequate Communication : Many parts of the areas are made up of difficult physical features. Such difficulties are compounded with lack of adequate communication systems in many areas, particularly during the rainy season. These difficulties restrict the movements not only for general public but also for the guardians of the wildlife sanctuaries and forest officials to take appropriate timely action and monitoring.

(3) Administrative Difficulties : Due to existence of several political entities, for example, into states or district councils within a state, as in Meghalaya, enforcement of comprehensive conservation programmes are not administratively feasible. Moreover, different authorities have different ways of looking and treating the problems of wildlife protection.

(4) Meat Loving People : By nature tribal are meat lovers in this area. Probably because of their closeness and involvement with nature, hunting and trappings remain habit with still a large section of population. Such habit may not necessarily be because of his love for meat but more so because of the pleasure and satisfaction he perhaps derives out of it. This is certainly one of the contributing factors in the depletion of wildlife.

(5) Jhumming : The agricultural habits of the tribals and the hilly terrains of this area are conducive only for 'slash and burn method' of cultivation or jhumming. But with increasing populations, how long can such practices be continued? Jhumming has and is drastically reducing the wildlife habitats. Jhum related fires are also seen to be destroying nearby forests in many areas.

(6) Political Unrest : Insurgency, Extremism, Political Unrest, Agitational Movements and similar activities in North-East certainly add to the dimensions of the problems of wildlife protection. For example, we read illegal killing of wild animals and cutting of trees (by the agitationists in the absence of forest / wildlife guards or some watchers, who can not go due to fear) in the Manas area of Assam.

(7) Inadequate Resources : Lack of sufficient trained man power and scientific information on our wild animals, etc. are other features of problems of wildlife protection.

Community Action

From the foregoing background, proposals may be made possible in the areas of community action and participation for effective enhancement of nature conservation and wildlife protection. Perhaps many of our village committees may already be executing some of these measures, thereby arresting environmental degradation and preserving nature. One such example is the tradition of protecting and regarding certain pockets or patches of forests 'sacred'. Theses traditions of prudence came to be accepted and practised perhaps on the basis of intimate experience and contact of the people with the working of nature, although conveniently or unknowingly related to religious beliefs, followed by social compulsions. Nevertheless, these were effective and praiseworthy ways of protecting the nature. Since our people are gifted with this traditional wisdom, the task is to manipulate this concept and recreate in the modern line. Some aspects of community action and participation in arresting environmental degradation and wildlife protection may include :

(1) Awareness, Information and Education

- to increase the community awareness of the different aspects of the problems and their dimensions, even through non-formal education,
- to seek information on various aspects of wildlife of your locality/state/ from the wildlife department or other agencies like Zoological Survey of India, and
- request for exhibits and displays in your local schools

or community halls which will provide additional avenues of awareness and education.

(2) Build-up Concerned Citizen/Community

- with the above steps you could be an informed citizen and assist your community to be an informed community,
- informed citizenry would simulate you to concerned citizenry and concerned community,
- concerned community and collective actions could do miracles in protecting nature and wildlife,
- exploit other avenues of resources to minimize pressure on forests, and
- appreciate the role and importance of wildlife for the well being and survival of mankind.

(3) Some Suggested Actions

a) Recall that in Meghalaya about 90 per cent of forest are either under the control of District Councils, Village Community or Private. This is where the importance of community action becomes more important in the context of rational use of forest resources, thereby enduring sustained food and shelter for different categories of wildlife. For instance, the village community may agree to extract only 1-2 headloads of firewood per family per week from the community forest and so on.

b) The spirit of re-creation of 'sacred groves' may be revived. Communities may afford protection particularly to vegetations growing in the serpentine river banks or streams. Even stretch of rivers / streams can be afforded similar protection. Such protected patches would attract varieties of organisms.

c) While demarcating non-forest lands like agricultural lands, the villagers may establish green belts or 'hedge' vegetations to be kept but undisturbed as far as possible. Such measures may help many a species of animals to recolonise areas under humanuse, from which these animals have once disappeared.

d) Concerned communities must try to convince their children and others not to do shooting or hunting of animals or birds, and poisoning of rivers. Assist them to derive pleasure in watching and observing the birds and animals around your community. On the other hand, the communities may formulate certain regulations and punish the violators.

e) Communities living near the reserved forests and Sanctuaries / National Parks have special role to play in the protection of wild animals. Once awareness is built up and the importance of wildlife is appreciated, such communities in the periphery of sanctuaries could act as protective forces and help the authorities in apprehending the poachers and intruders.

f) Communities can establish and demarcate certain areas for Community Fodder Farm in which fast growing quality plants / grasses can be grown. This would help in availability of continuous grazing lands (or check overgrazing) thereby reducing pressure on forests and competition with wildlife.

g) Concerned citizens / community can jointly voice against excess / unauthorized extraction of timbers and forest products by contractors, individuals or even at times by,the authorities in their own localities, thereby ensuring protection of animals in their own localities.

h) Even in the areas where the communities must depend on jhumming as in parts of Garo Hills, Ri-Bhoi areas and Jaintia Hills, jhum land can be carved out prudently by keeping continuous patches of forested tracts untouched. Such forest patches would serve as corridors for movement of wildlife as well as provide food and shelter, ensuring survival of both man and animals.

i) Due to their needs and activities (for fuel, building materials, agriculture, etc.) people contribute to environmental degradation and shrinkage of habitat for wild animals. Each year each community can work out common action plans to be carried out to protect resources around them. They can initiate dialogue and seek assistance from various regulatory bodies and related governmental departments.

j) Communities can always feed back their traditional wisdom and experience of nature conservation / wildlife protection so that our development planners can incorporate such specific wisdoms and experiences perhaps with scientific modifications while preparing developmental planning of the areas.

Concluding Remarks

Experts tell us that if current rate of environmental degradation continues, 25 per cent of the world's planet and animal species face extinction within the next 25-50 years. Therefore, at the grass-root village or community levels, we must arrest the changing patterns of landuse, loss of wild habitats, pollutions, overharvesting of species and so on. To achieve these, what we need is people's innovation, creativity and involvement with promotion of sustainable management practices in forestry, wildlife, agriculture, fisheries and allied fields alongwith transfer of know-how, technology to communities to enable them care their own local plants and animals. As many as possible, small or large, we must create zoological parks and botanical gardens or biocomplexes to save and maintain local species. Once the people appreciate the values nature offer to them, such goals would be easier to achieve in places like Meghalaya where the land belongs to the community and perhaps the communities would then spontaneously come up with offer of portions of their lands for conversion into parks/gardens. The need is to evolve comprehensive wildlife programmes that can help forge a 'new partnership ... between people and nature' (UNEP).

Acknowledgements

I am thankful to Mr. P.A. Thorose, Principal, St. Peter's School, Mr. T.C. Marak, IFS, Planning Officer, Forest Department, Government of Meghalaya and Mr. C. Budnah, IFS, DFO Wildlife, Khasi Hills for providing informations with useful suggestions.

14

Regeneration Potential of Tree Species in Meghalaya Forest

M. L. Khan
Department of Forestry,
North-Eastern Institute of Science and Technology, Itanagar

Introduction

The forest cover in Meghalaya is depleting at an alarming rate due to felling of trees for shifting cultivation, timber and fuel, and on account of developmental activities such as road construction, hydro-electric projects and new industrial set up. Shortening of jhum cycle due to population increase deprives of the jhum land of sufficient time required for vegetation reestablishment.

At lower elevation in the state, a matrix of broad-leaved forest was well stocked until last two decades. Establishment of a few plywood industries in this area has however, encouraged unplanned extraction of plywood species. Consequently, the matrix of forest is reduced to the scattered patches located in difficult and inaccessible areas. At upper elevation, semi-evergreen forest type vegetation forms the natural community, whereas pine constitutes the artificially raised forests on jhum lands. Most of the natural climax forest is cut and the productive land is utilized for cultivation, degrading it to different degrees based on the intermittent fallow period and frequency of fire.

According to an assessment made by the Survey of India, the closed canopy forest (crown density above 40%) accounts for 34.82 per cent, open canopy forest (10-40% crown density) 55.16 per cent and shrub vegetation accounts 1.04 per cent of the total forest area in Meghalaya (State of Forest Report, Ministry of Environment and Forests, Government of India, 1987). As indicated in the report, 16 per cent forests regenerate profusely, 26.4 per cent adequately, 49.2 per cent inadequately and 8.4 per cent forest do not show any regeneration. During field study, it has been observed that pine and some broad-leaved species regenerated naturally at many places, but because of fire hazards, most of the tender seedlings and sprouts were again wiped out and the area became almost barren supporting only stunted trees and grasses. Further, graziers do a lot of damage to the newly regenerated vegetation especially the fodder tree species. During winter when the annual grasses dried up, 'the sprouts, a successful mode of regeneration of broad-leaved tree species, are lopped heavily and at times the trees themselves lopped to death. This article is focused on suggestive measures to improve the tree regeneration.

Experimental Study

In a matrix of sub-tropical wet hill forest at Upper Shillong (1955 mean sea level) which is dominated by *Quercus dealbata, Q. griffithii, Manglietia insignis, Schima khasiana* and *Myrica esculenta,* and is exposed to frequent tree felling for timber and fuelwood, a mildly disturbed (about 7 ha.) and a highly disturbed (about 5 ha.) stands were marked. In each stands, tree seedling density was determined in June 1983 in randomly laid 20 quadrats of 20m x 20m. The tree seedlings were tagged and their survival was studied at bi-monthly intervals over one-year period (June 1983 to May 1984).

To analyse the effects of thickness and height of stump on emergence, survival and growth of sprouts of four dominant and economically important forest trees, *viz., Alnus nepalensis, Quercus dealbata, Q. griffithii* and *Schima khasiana* both in mildly and highly disturbed stands where tree felling occurred during December 1982, 10 quadrats of 20mX20m were randomly laid during June 1983. The stumps (sprouting and non-sprouting) were counted in each quadrat for density estimate and thickness of stumps of each species was measured. The stumps were then categorized into 5 diameter classes : less than or equal to 15, 15-30, 30-45, 45-60 and > 60cm. Stumps falling in the most common diameter range (15-45 cm.) were further demarcated in 4 height classes 5-10, 25-30, 45-50 and 65-70 cm. Ten stumps of each species in all diameter as well as height classes were separately tagged to study the survival and extension growth of sprouts at bi-monthly interval over a period of 2 years.

Mildly disturbed stand showed greater density of tree seedlings and their better survival compared to highly disturbed stand (Table-14.1). Reduced recruitment and survival of tree seedlings in the later stand is attributed to frequent fire which affects both soil seed bank and emergence of seedlings. Tree felling and grazing also adversely affect the seedling survival.

For all species, percentage of stumps that sprouted decreased with the increase in stump diameter (Table-14.2). Emergence, survival and growth of sprouts were greater on the stumps of medium diameter than the stumps of thinner or thicker diameter (Tables-14.3 & 14.4). With advancing tree age, the number of dormant buds which give rise to sprouts decreases and also the increase in thickness of bark offer much mechanical hindrance to sprout emergence. Sprouting and survival of sprouts and their growth increase with increase in stump height (Tables-14.5 & 14.6). In case of stumps of low height, the population of dormant buds happens to be low and this could cause decrease in sprouting on such stumps. Sprouting, survival and growth on low-cut stumps were adversely affected by their rapid decay. Further, inadequate food reserves in low-cut stumps may reduce the survival and growth of the sprouts emerging from such stumps.

Sprout survival was much higher in mildly disturbed stand as compared to highly disturbed stand (Tables-14.5 & 14.6). Frequent recurrence of disturbances caused greater damage to the sprouts in the latter stand.

Table-14.1 : Density of Survival of Tree Seedlings in Mildy and Highly Distributed Forest in Upper Shillong.

Tree Species	Density (ha.)	Survival (%)
Mildly Disturbed Stand		
Alnus nepalensis	90	40
Myrica esculenta	35	32
Pinus kesiya	500	60
Quercus dealbata	130	39
Q. griffithii	86	46
Schima khasiana	180	38
Highly Disturbed Stand		
A. nepalensis	30	18
M. esculenta	15	15
P. kesiya	208	32
Q. dealbata	60	10
Q. griffithii	50	8
S. khasiana	85	12

Table-14.2 :Mean Density (per ha.) and Sprouting Percentage of Stumps (in parentheses) of Four Tree Species in Five Diameter Classes in the First Growing Season after Six Months of Cutting in Two Forests Stand in Upper Shillong.

Species	Diameter Class (cm.)				
	< 15	15-30	30-45	45-60	>60
Mildly Disturbed Stand					
Alnus nepalensis	2	9	12	7	-
	(100)	(92)	(90)	(70)	
Quercus dealbata	3	12	10	8	-
	(70)	(68)	(60)	(35)	
Q. griffithii	2	10	15	6	-
	(70)	(55)	(40)	(25)	
Schima khasiana	7	10	20	9	5
	(100)	(86)	(85)	(63)	(15)
Highly Disturbed Stand					
A. nepalensis	4	20	26	9	5
	(100)	(90)	(90)	(70)	(62)
Q. dealbata	6	25	32	15	-
	(70)	(65)	(60)	(32)	
Q. griffithii	5	23	35	17	3
	(72)	(56)	(42)	(23)	(0)
S. khasiana	9	32	39	20	6
	(100)	(85)	(82)	(60)	(10)

Table-14.3 : Mean Sprout Number Per Stump in the First Growing Season after six Months of Cutting and Mean Height Per Sprout Attaining in Two Years Period (cm., parentheses) in Five Diameter Classes for Four Tree Species.

Species	Diameter Class (cm.)				
	<15	15-30	30-45	45-60	>60
Alnus nepalensis	7 (140)	10 (173)	14 (169)	12 (97)	4 (75)
Quercus dealbata	8 (92)	14 (125)	10 (120)	4 (58)	-
Q. griffithii	3 (72)	8 (96)	8 (74)	2 (44)	-
Schima khasiana	12 (154)	16 (169)	22 (168)	13 (97)	4 (74)

N.B. : Differences in mean sprout number and mean height between mildly and highly disturbed stands were insignificant.

Table-14.4 : Survival Rate (%) of the Sprout Emerged from the Stumps Classified in Five Diameter Classes in Two Forests Stands in Upper Shillong.

Species	Diameter Class (cm.)				
	< 15	15-30	30-45	45-60	>60
Mildly Disturbed Stand					
Alnus nepalensis	70	75	80	48	23
Quercus dealbata	68	78	79	10	-
Q. griffithii 35	48	70	8	-	
Schima khasiana	56	60	75	48	5
Highly Disturbed Stand					
A. nepalensis	40	43	44	23	10
Q. dealbata 20	25	30	5	-	
Q. griffithii 20	22	25	6	-	
S. khasiana 35	36	32	20	-	

Table-14.5 : Number of Sprouts Per Stump in the First Growing Season after Six Months of Cutting and Mean Height Per Sprout Attained in Two Years Period (cm., parentheses) as Related to Stump Height in Four Tree Species.

Species	Diameter Class (cm.)			
	5-10	25-30	45-50	6 5 - 7 0
Alnus nepalensis	4	6	10	12
	(144)	(168)	(188)	(205)
Quercus dealbata	7	10	11	15
	(96)	(133)	(153)	(172)
Q. griffithii 2	3	5	6	
	(82)	(90)	(105)	(123)
Schima khasiana	5	10	14	22
	(154)	(182)	(198)	(225)

N.B. : Differences in mean sprout number and mean height between mildly and highly disturbed stands are insignificant.

Table-14.6 : Survival Rate in (%) of the Sprouts Emerged from the Stumps Classified in Height Classes in Two Forests Stand at Upper Shillong.

Species	Diameter Class (cm.)			
	5-10	25-30	45-50	65-70
Mildly Disturbed Stand				
Alnus nepalensis	20	75	68	30
Quercus dealbata	22	83	75	45
Q. griffithii 15	65	60	30	
Schima khasiana	50	72	65	60
Highly Disturbed Stand				
A. nepalensis	8	46	34	5
Q. dealbata 10	42	35	20	
Q. griffithii 10	39	29	10	
S. khasiana 22	52	30	20	

Suggestive Measures

It was often noticed during field survey that pine and some broad-leaved tree species regenerated profusely in clear-cut forests. But, induced fire wiped out most of the tender seedlings and sprouts. It is, therefore, needed to protect potential regenerating sites from fire and other disturbances. Local people who are accustomed to cause such disturbances could be taken into confidence by communicating them the importance of regeneration process. Weed infestation must be checked. Poor natural regeneration sites should be aided with planted seedlings. Enrichment should be done preferably in gap areas created by selective removal of trees. Social forestry programme which aims to rehabilitate the degraded forest stands may play pivotal role in providing an umbrella for improving the natural regeneration of forests.

It is concluded that the forests in Meghalaya have a great deal of regeneration potential. If natural regeneration is allowed and supplemented with seedling plantings in those areas where this process is relatively poor, vast tracts of forest could be maintained and also the forest could be extended in peripheral degraded areas. Further, in such forests where tree felling is done by forest department regeneration couldbe maximised by appropriate combination of treatments which may ensure cutting of medium diameter trees at optimum height.

Subject Index

Afforestation 6,16,19,133.
Agricultural Development 81.
Agro-based 114.
Agronomy 8.
Alkali 5,6.
Anthropology 12.

Bio Diversity 36.
Biological Survey 12.
Biomass 29,31,142.
'Bun' Method 5.

Carbon dioxide 13.
Community Land 87,88,157,158,159.
Conservation 36,37,123,155.
Consumption 19.
Cropping Intensity 11,106.
-Pattern 11,69,100,105.
-Production 69.

Deforested 5,97,133,138.
Degradation 4,5,7,8,24,31,39,54,119,120,121,124,127,136,137.
Demarcation 6.
Drainage 7.
Dyke 24.
Dry 25.

Ecology 8,9,23,34,40.
Ecorestoration 5,9,10,11,13,23,24,34,37.

Ecosystem Degraded 23,29,33,35,41,130.

Egalitarian System 91.

Environment 8,9,11,119,120,121,124,127,131,133,147,148.

Erosion 5,15,16,17,77,98,128,138.

Forest Reserve 17,150.
-Conservation 18,19.
-Department 18.
-Sector 18.

Geohydrological 11.

Geomorphic Process 127,128.

Grasslands 24,27,28,29,30,35,38.

Grazing 19,39,59,60,163.

Green Manuring 6.

Growth 10.

Industrialisation 10.

Integrated 9,11.

Jhum 23,54,61,62,63,65,66,67,68,70,73,74,76,112,115,161.

Khasi Customs 90,92.

Landscape 24.

Land Ownership 79,80,81,92.

Landuse 8,55,64,81,91,93,95,96,100,101,102,110,111,112,115,121, 123,131,143.

Leaching 27,32,37.

Livestock 55,56,57,59,112.

Mammals 151.

Manuring 21.
Moisture 7,33.

Nitrogen 19,27.
Nutrient Accumulation 31,32,47.
-Contents 37,50.
-Dynamics 31.

Operation 11.
Organic Matter 33.
Oxygen 13.

Pasture 7,59.
Physiography 3, 105, 115.
Phytomass 37.
Plan 15.
Plant Nutrients 25.
-Growth 25.
Plantation 20,35.
Protection 17, 21, 39, 153, 154.
Pollution 10.

Quality 17.
Quantity 17.

Ravines 4, 5.
Recommendation 6.
Research Extension 77.
Resources 17,24,53,77,124,135,137,139,140,153.

Saline 5,6.

Scientific Community 10.
-Organisation 10,11.

Shifting Cultivation 53,54,55,63,71,76,77,78,92,98,103,111,122.

Sociology 8.

Soil Conservation 11,15,16.
-Erosion 60,121,124.
-Fertility 25.
-Profile 38.
-Texture 31.

Sub-tropical 25,162.

Spur 14.

Stumps 165,166.

Technology 7.

Tenency System 103,105,109.

Topography 33.

Tree Species 19,20,27,161.

Turfing 16.

Vegetation 24,27.

Wastelands 4,5,6.

Weather 33.

Weeding 16,20,21.

Wet 25.

Zoological Survey 12.